To Brian and Jacqueline

with every

Best Wish,

Peter Wilson

PURCHASE TWO KILTS

Purchase Two Kilts

by

PETER WILSON

The Memoir Club

© Peter Wilson 2001

First published in 2001 by
The Memoir Club
Whitworth Hall
Spennymoor
County Durham

British Library Cataloguing in
Publication Data.
A catalogue record for this book
is available from the
British Library.

ISBN: 1 84104 012 6

Typeset by George Wishart & Associates, Whitley Bay.
Printed by Bookcraft (Bath) Ltd.

Dedication

This book is dedicated to my wife, Bunny,
without whose love, support and encouragement
over the last fifty years this book could not have been
written, as there would have been no story to tell.

Contents

List of Illustrations

Foreword

AUTOBIOGRAPHY IS THE most delicate of the literary arts. It can be an essay in self-discovery; it can be a publicly available self-understanding; it can be the window of which the screen has been drawn back by the householder. I have no doubt that each of these is applicable in its own way to *Purchase Two Kilts*. What is evidently also true is that this is a story of both human interest and human development.

It is not yet the full 'seven stages of man' which Shakespeare distils with humour and pathos, but perhaps the tale, beginning with the childhood which Peter Wilson tells, does show him

> 'Full of wise saws and modern instances,
> And so he plays his part.'

The threads which hold each human life (and in this case two happy and mutually supportive human lives) together, are different and therefore weave different portraits.

Each reader will have particular interests to be prompted and for this one, the chapters on Wye College and the University of Edinburgh had particular resonance.

Stuart R. Sutherland, FBA, FRSE,
Principal and Vice Chancellor,
University of Edinburgh

Preface

THE IDEA OF WRITING this book came from three different sources. Firstly, my former colleague from Uganda, Willem Boshoff, kindly sent me a copy of his autobiography, *My African Notebook*, (176) which set me wondering if I would ever pluck up enough courage to produce something similar. Secondly, my farming friend Ben Coutts asked my advice about publishing his own autobiography, From *Bothy to Big Ben* (174), which he had submitted to several publishers without success. Whether or not I was able to help him I am unsure, but two years later his book was printed and to me it made very interesting reading. Lastly, and most important of all, in 1998 I received a letter 'out of the blue' from the Editor of the Memoir Club asking if I would agree to write an autobiography for publication by his Club. After some correspondence, and a meeting near Durham, I agreed to proceed and signed the contract.

It has always seemed to me to be a little pretentious to write an autobiography. I had assumed they should all be written by good and great people whose names are well known. My own career is very undistinguished. All that can be said about it is that nothing was planned! I served in Africa on the staff of one of the early Universities set up just after the War. I was the last British Professor of Agriculture at the University of the West Indies. I then spent nineteen years with what I whimsically call 'The small family firm' of Unilever, first as a Research Scientist, then as a Company Director and latterly as Chief Agricultural Adviser. Finally, I returned to academe to assume the Chair of Agriculture and Rural Economy at Edinburgh University, until I retired ten years ago.

Unplanned it may have been, but it has been an interesting and eventful career. If there is a key word that knits the whole story together, it is 'Change'. None of the institutions I have served has stood still and most have evolved, and are still evolving, very radically. I have seen these changes come and some of them go. It would seem that I am 'bad news' for any institution that yearns for stability, for my presence seems to exert the opposite effect. Perhaps I have been party to the reasons for change wherever I have pitched my tent.

As a dedicated agriculturist, with a passionate belief in the importance and long-term future of farming, it is salutary to ask what effect I have had on agricultural matters. The answer, I suspect, is very little. The causes of change have been beyond my ability to influence for good or ill. The only footsteps which I may have left on the sands of time have been my papers. I was advised by Sir John Hammond always to publish papers of two sorts: those of some interest and possible significance to the scientific world and those of more transient relevance to those connected, in one way or another, with farming. I have followed this advice and the 170 or so papers I have written, as sole or co-author, are all listed. I do not expect the reader to chase them up in a library (although full references are given for each paper), but it is possible that some specialist readers will be interested to see what I wrote about their subject thirty, forty or fifty years ago. Obviously some papers may have been better left unwritten, but most have stood the test of time and I would not wish to alter many of their conclusions.

Farming is going through a rough period and I try to give some of the reasons in this book. I am confident it will emerge from this bad patch just as it has done down through the centuries. At the end of the day, man needs to be fed; in our present stage of evolution most of this food needs to come from farms. If there are heroes in this book, they are the farmers. I salute them all and only wish that Society would do the same, as it did during the hard times of War.

Acknowledgements

FIRST AND FOREMOST, I wish to thank my wife, Bunny, who has not only endured for over a year my preoccupation with the writing of this book, but who has given me unstinted help in reading and correcting the several drafts through which the manuscript has passed.

I wish to express my thanks to Lady Janet Balfour for reading the manuscript and giving me much valued useful professional criticism. Her encouragement to persevere when other matters intruded upon my time was especially helpful.

I also wish to thank the following for help with various sections of the book:

Dr Lesley Campbell for regular help with overcoming my reticence to master my word processor properly, and for ensuring that main drives, floppy discs and hard copy were all running in perfect parallel.

Mr F.H.G. Percy, for help with chapter 2.

Mr T.R. Houghton, for help with chapter 8.

Mr J.A. Inverarity, CBE, for help with chapters 12 and 14.

Mr D. Osler, for help with chapter 15.

I also wish to acknowledge my thanks to the following, for permission to reproduce photographs:

Plates 6, 7, 44 – H.M. King of Croydon

Plate 9 – P.G. Bartlett, Studio 8, Ludlow, Salop

Plates 10 and 44 – Museum of Rural Life, Reading, Berks.

Plate 11 – G. Thompson, Waddon, Croydon

Plate 13 – Hulton Getty, Woodfield Road, London W9 2BA

Plate 24 – J.S. da Costa, Kampala, Uganda

Plates 30 and 31 – Julse Williams, Heswall, Cheshire

Plate 32 – Tom Kroeze, Rotterdam

Plate 34 – Derek Rowe, 1-4, Pope Street, London SE1 3PH

Plate 35 – Viewpoint Photo Services, Warnborough, Basingstoke

Plate 38 – Photo Express, 7 Melville Terrace, Edinburgh

Plate 41 – Hannah's, North Berwick, E. Lothian

Plate 46 – Michael Bryant, 11 Midland Road, Bedford

Plate 52 – Spicer Halfield, Stirling

Plate 53 – Surrey Press, Croydon
Plate 54 – Orushke Photographers, Jerusalem
Plate 56 – Archie Handford, Croydon
Plate 59 – *Journal of Animal Science*, 1995, vol. 60, p. 1

Although every attempt has been made to contact the photographers listed above, due to passage of time and disturbances and civil disruption overseas, some have not replied to requests for reproduction of their material. I nevertheless duly acknowledge their work, even though request letters have been unanswered for these and other reasons.

Lastly my thanks to my several 'good friends and mentors' who have been the subject of the 19th chapter of the book. Some, alas, are no longer with us and the others may find their entries in the book surprising, but I hope not too embarrassing. I owe them a debt of gratitude, not in helping me to write this book but in guiding my career in a manner that enabled the book to be written.

CHAPTER 1

Early Days

'Ah, yes. I remember it well!'
Maurice Chevalier

IT IS DIFFICULT TO KNOW how far back our real memory can stretch. Much must surely depend on the reinforcement of far-off events by family anecdotes and old photographs. That said, there are certain early incidents in my life which I am confident I recall from memory, without any assistance from other sources.

My first clear memory is of sitting on a rug under an apple tree in the garden of my first home – a council house in Goddard Road, Beckenham, Kent. My parents were keen gardeners and I inherited a love of creating order out of chaos from them My family moved from Beckenham when I was four, but the image in my mind is of crawling from one side of the rug to the other to gather, and put in a wicker basket, the apples which my twin brothers were picking and throwing down from the tree. I was still at the crawling, rather than the more mobile toddling, stage, which probably puts my age at the time at about 18 months. I have revisited the house and the garden with the apple tree is still there – a gnarled and somewhat unproductive fruit tree, which has escaped the gardeners' axe and is now well past its normal lifespan. The blanket has long since disappeared!

The next recollection from those days is remembered with much less enjoyment. The house had a coal cellar underneath the stairs, fairly close to the front door. It supplied fuel for the open fires which were installed in most of the rooms and was usually quite full. Why the coal cellar was placed in such a strange position I do not know – it has long since gone – but obviously the disadvantage of such a dirty place so near to the living rooms of the house was offset by the advantage of not having to go outside the house into the cold to fetch the next bucketful of coal. Whatever the reason, this coal cellar terrified me, for on one occasion I was punished by my mother by being shut up in this dark and dreary dungeon for what seemed like an age. The crime must have been serious, although I cannot remember what it was, but the effect of the incarceration in the pitch-black cellar had a long-lasting effect on me. I am still somewhat afraid of coalmines and of

1

1. The author with his mother and twin brothers, Arthur and Donald, 1928.

being shut up underground and this fear extends to caves, dungeons and submarines. I could never be a potholer or a submariner, although I can now enter large caves and visit small pits and dungeons in castles without difficulty. I often wonder whether my fear of the danger of being confined underground was inborn and surfaced with the punishment in the cellar, or whether the hour or so in the 'dark black room' was the trigger for the mild claustrophobia from which I have suffered ever since. Which came first – the chicken or the egg? One will never know.

My next home was in Sanderstead, Surrey. We moved there from Beckenham in 1932. I do not remember the actual move, but I recall vividly the pride of my parents in living in a brand new house on a large estate, with several modern gadgets. One was a movable kitchen table which slid up into the wall; in its place an ironing board would descend from its small cupboard at the push of a button. Central heating was provided by an 'Ideal' boiler in the 'breakfast room', which was coke-fired, with the fuel kept – thank heavens! – in a bunker outside the back door. These 'modern features' were constantly being exhibited to various friends and relations who descended upon us to see this up-to-date, fully equipped, new semi-detached house and I joined in my parents' pride and became a keen 'demonstrator' of the various pieces of household equipment. One of the features, ancient rather than modern, was a boiler in which clothes could be washed or in the summer, fruit could be 'sterilised' in preparation for storage in Kellner jars. I had the job of helping to prepare plums, damsons, apricots and greengages for their preservation in this way and I can still see my father checking the rubber seals round each cooled jar for any possible air leaks, which, I was told, could cause the fruit to become 'poisonous'. Another of my jobs was to grind down large blocks of solid salt for domestic use and also to cut up long bars of 'Lifebuoy' soap into more suitably sized tablets.

As the house was new, the garden was not made up and was merely a part of a sloping field which had been down to pasture for many years. My father and my older twin brothers created a very pretty, if conventional, garden from this rough grazing, which entailed moving many tons of soil from the back of the house further up the long, thin building plot. I can recall trying, no doubt not very successfully, to assist them with this work. The object was to level the ground, forming a small lawn and flower border near the house, with a curved flight of steps going up to the larger lawn beyond. This was flanked by a flower border on one side and a row of trees on the other. I asked to be given a piece of 'garden' for my own use and was duly allocated a small area underneath the row of trees at the furthest end of the garden. Here I created my first rockery, quite a complex affair with small winding paths, which I planted down to grass. It was flanked by rockery bricks, bits of concrete from the rubble left by the builders, with a small flower garden on the top. This was quite effective and much admired, and I spent many hours moving stones and planting new flowers, so that this rockery garden became quite a prized feature of our new home. The rockery was still 'my special garden' when I left home to go to University in 1946.

The tree adjacent to my rockery was a tall ash with several slender trunks, and I used to climb into its branches to read a book or to admire my handiwork below. I did not make a tree-house (I doubt if this would have been allowed), but I knew every twig of that ash tree and must have spent many hours curled up on one of the natural tree-platforms formed by adjacent interlocking branches. One fateful day a branch gave way beneath me and I fell some 25 feet or so on to my rockery. By sheer good fortune, I fell neatly on to one of my newly created grass paths and therefore escaped serious injury, with only a few bruises to show for the escapade. I took greater care in selecting my ash tree perches from then on and have always remembered my father's important advice that, when climbing trees, either two feet and one hand should be secure or two hands and one foot. When I fell onto the rockery, all my four limbs were, I fear, 'at ease'.

I revisited our Sanderstead home in 1998. I was shown round the old garden, but it was not at all as I remembered it. The small garden with the border and flight of stone steps nearest to the house had somehow 'merged' with the lawn beyond, to form a single undulating grassed area. My rockery

2. The author's rockery garden at his home in Sanderstead, 1937.

had disappeared and the large 'lawn' was filled in with so many trees and bushes that it was not possible to discern the original 'hedge' at the side of the field, which had once formed the garden's boundary line. However the ash tree is still there, now much higher than before, with a missing section near the top due to the loss of the branch which I broke whilst immersed in one of my books.

During my childhood it was the family custom to take our summer holidays at Frinton-on-Sea in Essex. There were so many 'holidays by the seaside' at Frinton that I cannot recall the actual number of years involved. About a month before the holiday, my mother would open a large cabin trunk on the landing near my bedroom. Each day some garments or other objects would be ceremoniously placed in the trunk – perhaps my swimming costume (trunks were not then invented!), perhaps some casual wear belonging to my father or mother. I took a great interest in this sequential packing of the luggage, anxious to see that nothing of importance had been left out, and tracking the time before the holiday started by seeing how near to full the trunk was getting. Completion of this interesting task had to take place about a week before the holiday started, when my father would ask the local stationmaster to collect the trunk and send it 'Luggage in Advance' to Frinton-on-Sea. This seemed to me to be a hazardous operation – what if the luggage never arrived? How could we have a holiday without our clothes and other paraphernalia? Would we have to turn round and come home again, holiday abandoned? The seven days between the despatch of the trunk and the start of the holiday were worrying days indeed, and I slept a little more lightly during that anxious week.

Come the day of the holiday, the great excitement of the year took place: a taxi came to take us, and our minor hand luggage, to Liverpool Street Station, a journey taking about one half hour. Holidays were the only times we ever travelled by taxi. My worry about the safe arrival of the cabin trunk was now overtaken by an even greater worry: would we get to Liverpool Street Station before the train left? In fact, there was little chance of missing the train as my father always allowed the best part of an hour for any unforeseen eventualities, and we always had to stand or sit around on the platform waiting for the engine to be backed on. My father would leave the family party and go off to buy a newspaper or a magazine and, if we were lucky, some sweets to eat on the train. So my next major concern was whether he would remember on which platform he had left us. What would we do if the train started its journey without him? He had all the tickets; would we be allowed on the train minus father and minus tickets? The thought was a terrifying one.

Thankfully, however, year after year, everything went according to plan. In those days the trains ran strictly to schedule (my father used to check his watch by the passing of the Brighton Belle each morning through South Croydon Station) and the generous hour of grace always scheduled into our itinerary meant that there was no chance of missing trains or losing members of the family at Liverpool Street Station. Also, the impeccable organisation of the arrangements for 'Luggage in Advance' always meant that our trunk was ready and waiting at Frinton-on-Sea when we arrived.

In the latter days at Frinton, when I must have been 8 or 9 years old, I looked forward to a part of the holiday which to me was a highlight but, to my two twin brothers, a waste of time. The station at Frinton-on-Sea was last but one on the line to Walton-on-the-Naze and at the end of the station platform there was a manually operated level-crossing. This was closed just before the train arrived at the station and opened again, usually by the signalman − whose name was Fred − after the train had passed through. Fred was a small, round individual, usually rather short of breath as well as short in stature. He smoked a pipe and always had his standard issue railway uniform cap perched on his head at a rakish angle. He was a kindly man with twinkling blue eyes and he seemed to have a soft spot for me. I used to be allowed to 'help' Fred unlock and move the level-crossing gates. No doubt a highly irregular manoeuvre, which nowadays would be forbidden by the Health and Safety at Work Executive, and I realised it was a great privilege to be actually walking across the railway lines before the gates were 'officially'

3. The author with his father, travelling to Frinton-on-Sea, 1933.

fully open. Over the years my family and I got to know Fred the friendly signalman very well and he generously allowed me to 'assist' him in his duties. I even got invited up into the signal box and had the railway telegraph system explained to me, but I was never allowed to operate the levers which controlled the signals or the points. (I probably wouldn't have been strong enough anyway!). My holiday 'duties' on the level-crossing gates were something I looked forward to from one year to the next. It always pleased me when we climbed down from our train to be greeted by Fred with his friendly smile and the words, 'Good to see you back, Mr Peter. Will you be giving me a hand with them 'eavy gates again this year?'

Two other events are engraved on my memory from our happy summer holidays at Frinton-on-Sea. The first was the daily delivery of milk by a milkman driving a horse and cart. The horse, known as Blondie (a strange name for a dark chestnut gelding!), was partial to clover, which grew in abundance on the greensward between our holiday house and the sandy beach for which Frinton is well known. Last thing every evening I walked down to the greensward, picked a large armful of white and purple clover, and waited each morning for the punctual arrival of Blondie pulling the milk cart. Blondie, like Fred at the railway station, seemed to outlive our many years of holidaying at Frinton and I can only assume that the large amounts of clover I fed him did not give him colic, for legumes are somewhat strong meat when fed in excessive amounts to equines.

The second incident was a much more frightening experience, which had a similar long-term effect on me as my punishment in the Beckenham coal cellar. Early one morning, as we were getting ready for our daily visit to the beach, several fire engines roared past our house with bells ringing and sirens sounding. They stopped a couple of streets away, outside a furniture store. My father and I ran down the street to see what was going on and were confronted by the awesome sight of flames leaping from the roof of the building, which was well and truly alight. Staff had arrived to try to salvage what little they could of the furniture and I remember vividly seeing a three-piece suite pulled out and men dashing back inside to retrieve other items, only to be quickly driven out again by the flames. The fire-fighters' attention was directed towards saving the surrounding properties and the furniture store was simply left to burn itself out. After about half an hour the roof crashed in, bringing with it part of the outside wall adjacent to the road, and several fire-fighters had to jump for their lives to avoid the falling debris.

All this excitement was a little too much for me. I remember repeatedly asking my father whether anyone was trapped inside, but of course he did not know. I had nightmares for several years after this incident and insisted that the landing light in our house was left on at night and the door to my bedroom left ajar. I was terrified of the dreadful things that might happen in the dark and I wanted to be able to see my escape route clearly, should trouble brew up. This fear of the dark and of fire continued until the War years, when other more terrifying things occupied my attention. I think the Frinton fire must have happened when I was about four, and my nightmares continued until I was about nine or ten. When I am told that horrific TV and cinema films make little or no impression on young people, I wonder on what basis these so-called authoritative statements are made. In my opinion, the young are not only capable of experiencing long-term fear and distress as a result of seeing frightening sights, but could well have their normal behaviour patterns warped or distorted as a consequence.

In between these exciting events, Frinton holidays were for me sheer joy and delight. Long days were spent digging sandcastles on the beach, hours spent sailing a small boat in rock pools, shrimping with Don and Arthur (my two older brothers), playing ball games with other children – activities all full of great enjoyment. On Sundays we would walk through the fields from Frinton to Walton-on-the-Naze for evening service at the parish church. I can recall the blissful pleasure of walking along the footpaths between fields of corn bejewelled with poppies and cornflowers with skylarks singing overhead. The church had a peal of bells rung by a team of keen bell-ringers, so for most of the half-hour journey we were enticed on our way by the melody of chiming bells gently calling through the trees. I enjoyed the hymn singing and the lessons, but remember that the sermons were always long and boring and seemed to bear little relationship to the praise and worship of God. I recall after one long sermon (I think they averaged well over half an hour) saying to my mother: 'That wasn't a very good vicar – he never mentioned God or Jesus once!' But all was forgiven during the trek back home, when the shadows were long and mysterious and the air was silent and still. Sunday evensong meant an extra late night; late nights were a rare treat when normal bed-time was about 7.30, with an extra quarter of an hour being added every year as I grew older. Doubtless, as is always the case in the UK, there were wet days on holiday as well as fine, but I can only remember the warmth and the sunshine, the fields of golden corn waving gently in the breeze, ripening ready for harvest, and the joys of making sandcastles and paddling in the warm, shallow water.

Holidays came and went all too quickly and in between was 'real' life. School and housework; homework and school. My first school was a sort of pre-prep school class, run by a rather fearsome spinster who taught a group of about twelve children in her sitting room. I was told I had to 'sit an exam' before I would be accepted. I was four years old at the time, and I remember the 'examination' well. The schoolmistress – name long since forgotten – produced a box of carved wooden animals, representing Noah's Ark or a zoo collection. As she picked up each animal in turn, I had to give it its correct name. I had always been interested in animals and so had no difficulty in getting all, or most, of the names right. So I passed my first 'examination' and was admitted to the school! All something of a farce, since no one was ever kept out of the school because of ignorance of a sufficient number of animal names!

After a year in this pre-prep establishment, I went to a 'proper' prep school, run by two excellent, if somewhat elderly, tall, lean spinsters, the Misses Staunton and Sims. Their school was fifteen minutes walk from my house and, after being taken safely there on the first day or two of term, I was left to make my own way to 'Littleheath Prep School'. This establishment was exceedingly well run and I enjoyed the experience and learnt a lot. We played team games, rounders and football, we were taught to 'read' music (a sort of dot-and-dash score suitable for a percussion band). We were taught the three R's with rigour and in a no-nonsense manner; we were shown how to make wicker baskets and knit woollen squares for patchwork quilts and – greatest joy of all – we were given poetry to learn by heart. I can still (almost) recite the 'Pied Piper of Hamelin' from beginning to end, although I tend to get confused about the description of 'Brown rats, black rats . . .' and I still have a great affection for 'Tiger, Tiger, burning bright, in the forests of the night . . .' which I thought was just about the finest poem ever written.

During the walk to school I was on the look-out for discarded cigarette packets on the pavement, which I examined carefully in case one contained a collectable cigarette card. I built up a collection of several dozen albums of these cards, which, had they been kept over the years, I understand are now regarded as collectors' items. One completed the collection by trading duplicates in the school playground. These cards were little gems; usually 50 were required to complete an album and I am surprised that the practice of printing such sets of cards has not continued; it must have been a major boost to sales of the brand with the best cards. W.H. Wills, I recall, produced very good cards but Churchmans were regarded by the experts as being the best of the bunch.

I attended Littleheath Prep. School until I was ten, when I had to take a 'real' entrance exam to get into Whitgift School, one of the two schools administered by the Whitgift Foundation in Croydon, Surrey. My transition from prep school to Whitgift Junior School marked a major milestone in my life. I left behind the strong feminine influence of my mother and the two spinster school mistresses and entered the man's world of a day public school, taught by masters and ruled by prefects. I went to Whitgift the year before the outbreak of the Second World War.

My final early childhood memory is of having a long and very serious conversation with my grandmother, who lived with us in Sanderstead, about the prospects of war being declared. The year must have been 1938 and my grandmother solemnly told me that she was quite certain war would come and, when it did, it would be long and destructive. However she believed that 'right would prevail', as it was deemed to have done in the 1914-1918 First World War, but she foresaw that the battles to be fought would be long and bloody. She was proved right on most counts and my grandmother went up in my estimation as a result of these accurate predictions. My own parents fudged the question, expressing the vain hope that 'Hitler would see sense' and that war was not, therefore, inevitable.

However, Hitler did not see sense. The clouds of war gathered around us, and finally burst.

CHAPTER 2

Whitgift School

*'I have had playmates, I have had companions
In my days of childhood, in my joyful school days'*

Charles Lamb

WHITGIFT SCHOOL, Haling Park, South Croydon, was founded by Archbishop Whitgift in 1599. Whitgift endowed his foundation at the end of the Elizabethan era, following the founding of Westminster School in 1561, the opening of Rugby School in 1567 and Uppingham in 1584. Much later the educational part of the Foundation was divided into two, when Whitgift School moved to Haling Park (in South Croydon) in 1931, leaving the former site in North End, Croydon to continue to be used by an associated school, Whitgift Middle.

Admission to the school was by means of a competitive entrance examination. The proportion passing the exam was about 50 per cent for boys resident in the Borough of Croydon but only 20 per cent for non-Croydon-based boys like me. Twelve prestigious 'Victoria' scholarships were awarded each year, but preference was given to boys living in Croydon, so I was not eligible for one of these awards, which covered the cost of the school fees. In 1939 these fees were ten guineas a term, which today seems paltry, but to my parents thirty guineas a year was deemed to be a lot of money, especially considering that uniform, sports gear and books had to be bought in addition.

Although I passed the dreaded Entrance Examination, I did not do as well as my parents had hoped, partly because I was not good enough and partly because I was not taught Classics at my Prep. School, although Miss Sims did try to teach me some elementary Latin in the term before I sat the Entrance Examination. Part of the Latin examination was an exercise in scansion, an educational accomplishment apparently unknown to Miss Sims. Confronted with a question paper that demanded that I 'Scan the following lines', I read the paper carefully several times and, being ignorant of what was expected of me, I wrote in my best handwriting 'I scanned the lines carefully three times!' As a consequence of this gaffe, I was picked upon by the Latin master for the whole of my first school year. He would announce to the assembled class: 'I see we are honoured by the company of Wilson, that well-known scanner of Latin verse!'

My school days were happy and I look back on them with pleasure. This is perhaps surprising for two reasons. First, I attended the school for the duration of the Second World War and, in more ways than one, these were dangerous and troubled times. Second, I did not excel either academically or in the two formal school sports, rugby and cricket. My ranking in form was normally somewhere in the top ten out of a class of 30 and in games I only played for my House, Mason's, and never in any of the school teams. It was not until I entered the sixth form and concentrated on the sciences that I showed any academic promise, somewhat to the surprise of my science masters. But there were three activities which I pursued with vigour and it was these that allowed me to hold my own amongst my peers.

First and foremost, I was a keen and successful member of the school's Officers Training Corps (OTC). This being war-time, the OTC clearly had more prominence than it would have enjoyed in more peaceful days and I rose rapidly through the ranks of Cadet, Corporal, Sergeant, Colour Sergeant and (in my last month at school), Company Sergeant Major. I went to University when only 17 and had I been able to stay on an extra year at school, I might have risen to the highest rank obtainable, which was that of Under Officer. The boy who was level pegging or slightly ahead of me throughout my days in the OTC was a brilliant soldier-in-the-making, Colin Mitchell. He became an Under Officer as I left school to go to University and he eventually made a big name for himself as 'Mad Mitch', the well-known Colonel who played an outstanding part in the Aden campaign.

Although most of the instruction of the OTC cadets was given by the cadet NCOs (Non-Commissioned Officers) and WOs (Warrant Officers), the Corps was under the command of a number of masters, most of whom had seen war service. They kept a benevolent eye on things and were responsible for promoting the cadets to the various non-commissioned ranks. The masters gave up a lot of time to their duties and were on parade, in uniform, at least once and often several times a week. They accompanied the cadets on Field Days and also helped to train the shooting eight, which entered the public school shooting competitions at Bisley.

Three of these officers were especially respected by the cadets and they followed the boys' post-school careers for many years, sometimes throughout the whole of their lives. The first was Mr Robinson, the woodwork master, who, in a kind and gentle fashion, kept a fatherly eye on things, often without the boys realising how much their progress in the OTC was being carefully followed and monitored. Mr Robinson lived near to me in

4. The author as a Corporal in the Whitgift Officers' Training Corps, 1943.

Sanderstead and, when he realised that I was at something of a disadvantage in not having a bicycle to get to school during the War years, he kindly gave me his son's bicycle, which was no longer being used. The second was Mr Morris Thomas, the chief Chemistry master, who was always prepared to give boys struggling somewhat with his subject that little bit of extra out-of-school coaching. Morris was a very keen Freemason and played a prominent part in the Adeste Lodge, which was a closed Lodge catering solely for Old Boys from the School. I joined this Lodge on my return from my overseas career, in time becoming its Worshipful Master, and have kept in close touch with Morris Thomas for over fifty years, even when the Adeste Lodge eventually closed. This was because of lack of numbers on the one hand and the difficulty in finding a suitable Masonic Hall in which to meet on the other.

The third of these outstanding masters was Freddie Percy, who returned to the School from the army in January, 1946. Freddie became the School Archivist and took a great interest in the history of the school, which he researched in detail over many years publishing several books and papers on the subject (175). Freddie also acted as the Old Whitgiftian correspondent, responsible for the OWA section in the School Magazine and, as a consequence, he was by far and away the most knowledgeable person about the achievements of the Old Boys. He was a keen rugby footballer and captained the Old Boys' Rugby Football Club just before the War from 1938-39. He has played a prominent part in the Club since returning to the School after the War.

It is perhaps sad that the sterling worth of masters such as these is usually appreciated by the boys only after they have left school. During one's school days there is a tendency to take the staff for granted. It is accepted, indeed expected, that they are good teachers, but the time they devote to helping the boys in so many ways is not fully appreciated until much later. They have chosen a teaching career because they have a flair for it and genuinely like to see boys do well, on and off the field, in and out of the classroom and throughout their subsequent careers. Neither the decrees of Ministers of Education nor the strident voices of teachers' trade union leaders really do very much to improve the quality of teaching. The majority of the staff at the school during my time were scholars in their own right. They enjoyed passing on their knowledge to others and their constant wish was simply to see the boys do well and make their own unique contribution to society during the rest of their lives.

The other two activities which occupied my time were less traditional and perhaps illustrative of a slightly rebellious streak in my nature. School activities were very circumscribed. The team games were cricket and rugby, the uniformed extra-mural activities were the OTC and the Boy Scouts (very much considered the 'soft option' of the two) and the only individual game played (apart from athletics) was a variant of squash racquets, or fives, with the hand being used instead of a racquet to play the ball. I was interested in tennis, although not particularly good at the game, and I formed an unofficial School Tennis Club, which had the audacity to challenge the Croydon High School for Girls to a summer tennis match. This was not regarded kindly. I was interviewed by the Headmaster, Mr Ellis, who informed me dryly that only 'official' matches could be played and that these had to be against public schools in membership of the Head Masters' Conference. As the Croydon High School for Girls did not and could not qualify under these criteria, it followed that officially we could not play them – and certainly not at tennis! However, we did play them, unofficially and at the weekend, and were soundly beaten by the girls, who were coached in tennis by their games mistress as one of their official sports. I subsequently learnt that Mr Ellis was himself a very accomplished tennis player, so it must have been with his tongue in his cheek that he informed me that 'tennis really isn't one of the games that we play at this school'.

5. St. John's Youth Fellowship YHA Group, 1946. Seated, left to right: John Soper; Betty Finch; John Cannon; The Author; Andrew Harris; Margaret Herbert; Derek Holdway.

The third activity was to form a Youth Hostels Association (YHA) Group and promote organised cycling trips in the Easter and Summer holidays. This was not exactly frowned upon, as was playing tennis, but it was regarded as a bit odd, since new initiatives were not particularly welcomed in the 1940s. School life was very traditional in those days and innovations few and far between. However, half a dozen or so quite demanding Youth Hostel tours were organised by me and enjoyed by the few who took part in them. I am pleased to note that holiday trips of this nature are now a fully recognised activity of the school.

Whitgift School now has a roll of over one thousand boys, but in my time the number varied between 650 and 750. When the bombing campaign started in 1940, the educational authorities thought that they should give parents the choice of allowing their boys to continue as day boys (often in very difficult war-time conditions, when transport arrangements were being constantly disrupted by the Luftwaffe) or sending their children to a boarding school in Canada where it was assumed, they would be safe from the dangers of war. This must have been a very difficult decision for parents to take. I was adamant that I should stay put, and mercifully my parents agreed. Only about 30 or so parents opted for the Canadian boarding school offer and the children concerned were duly assembled a month later to board a ship bound for the St. Lawrence Seaway. A third of the way across the Atlantic the ship was torpedoed and all the 30 boys were lost, together with several hundred other passengers and crew. The agony of the bereaved parents was probably not eased by the knowledge that, of the 700 or so boys who remained in England, less than half a dozen lost their lives as a result of the blitz, although many more were injured and returned to school a term or so later to continue their studies. The reaction of the boys was fatalistic, though tinged with shock. Many had lost elder brothers or fathers in the War and had come to terms with sudden and untimely death. However many, like me, had pleaded with their parents to let them remain at home during the War and to them it was a relief to know that the decision to stay had turned out to be correct. But it was a very sombre Assembly at school when the Headmaster announced the loss of the ship and read out the names of the boys who had perished in the Atlantic Ocean.

School discipline was strict, although never oppressively so. Caning was practised but was infrequent. The average boy might expect to be caned by a master or a prefect about two or three times during his school career. The school prefects were responsible for discipline in the corridors, dining room, morning assembly and in the playgrounds. Masters were employed to teach; it was not their job to check on whether boys were correctly dressed or guilty of 'Going-Up-The-Down-Stairs' (GUDS) or 'Running-In-The-Corridor' (RIC). The prefects were deployed, by the Head Prefect, at strategic points in the School before morning assembly, during lunch and after school hours. Pupils committing any one of the half dozen or so 'crimes' were summoned to attend the weekly 'Prefects' Court'. The Head Prefect would hear the evidence and the excuses and then, after conferring with his colleagues, pronounce judgement. Punishments ranged from a severe ticking off, to the imposition of an essay or a number of 'lines' or, for repeated offences, up to six strokes of the cane. Boys had the right of appeal against caning to the Head Master, who also had the right to increase the sentence if he thought the punishment was too light.

I managed to get through school fairly lightly, being caned only twice by masters and once by a prefect during the course of eight years. In my experience, these punishments were meted out in an impartial manner without any vindictiveness. Discipline was well maintained and the boys were taught to think of the consequences of their actions and to accept the results of their misdemeanours. One particular master, Mr Fisher, our Geography teacher, had a cane permanently displayed beside his desk. He called the cane 'Archibald' and if a boy was thought to be out of line, he was threatened with the words, 'I see that Archibald will be needed soon to dust the seat of your pants!' This was no idle threat, although the cane was used only on relatively rare occasions; for all that, we were all in dread of the power of Archibald to inflict pain and suffering.

Charlie Fisher was a small, dapper man, with a walking habit reminiscent of Charlie Chaplin. He carried a silver-topped swagger-cane and always came to school each morning immaculately dressed in a prim suit, with well pressed creases in his trousers and a colourful waistcoat. The crowning glory of this well-turned-out character was that he invariably wore a fresh rose in his buttonhole each morning. Charlie Fisher was proud of his ability to sport a new bloom every day of the year and would sniff the scent from his rose from time to time. On one occasion a boy cheekily asked whether he might be allowed to sniff the nosegay. 'That would never do,' said Charlie Fisher. If all you boys sniffed my rose every day, there would not be enough scent left for me!' The logic of this argument always escaped us.

Charlie Fisher taught economic geography by rote. Every important town in the British Isles was associated with one or more important industries. Thus Kidderminster equalled carpets and when Charlie barked out the name 'Kidderminster', we were all expected to reply in chorus 'carpets'. Dundee, I recall, was famed for Jam, Jute and Journals. One boy, whose grandparents came from Dundee, thought he would be clever and instead of reciting the trilogy as expected, he embellished it with the words 'Raspberry jam, jute and journals!'. Charlie Fisher thought this most presumptuous and thundered that Dundee was renowned for making all sorts of jams. 'But,' exclaimed the boy, 'the Tay Valley grows a lot of raspberries so it follows that all or most of the Dundee jam must be raspberry jam!' 'Nonsense!' was the reply, 'In Dundee they know better than to make only one jam. If they can't grow different fruits, they can import them from elsewhere. Dundee is famed for jam – all jams – not merely raspberry jam!'

Charlie's lessons were always reinforced with numerous visual aids in the form of maps. Charlie loved his maps and some of his delight in the product of the cartographer's art must have spilled over, as I acquired a deep affection for maps from those early days. Over several years I amassed, with my meagre pocket money, a fairly complete collection of Bartholomew's half-inch-to-the-mile maps of the British Isles. The contours were coloured in for easy interpretation and the paper maps were cut and mounted on canvas, so that one never had the trouble in common with modern maps of having an important road junction or village name obliterated by a torn crease. I used these half-inch-to-the-mile maps for many years and still have most of the collection today, although they are now outdated by new developments, such as the creation of motorways and urban encroachment into the countryside.

From time to time as the War dragged on, there were staff shortages as the younger masters were progressively called up for war service, or volunteered for the armed forces. When a class could not be taken by a master, as none was available, a prefect would be sent for and he would either teach the class in question, if he was knowledgeable in the subject, or keep the form quiet whilst they read from 'set books' prescribed for the purpose. Prefects, in fact, made good teachers and their classes were much appreciated by the boys. I was made a school prefect in my last two years and I taught a dozen or so classes and enjoyed the experience. I only taught science classes and, as I was then reading science at an advanced level, I knew what I was talking about and the boys responded accordingly.

When, many years later, I was appointed to the Board of Governors of a Comprehensive School and asked to take part in the selection panel for a new head teacher, I was surprised at the answer I received to the question, 'What would you do if the staff were progressively struck down with 'flu, so that there were insufficient to teach all the classes?' The correct answer was 'I would send the classes home sequentially, starting with the most junior class and ending up with the examination classes'. 'But', I asked, 'Why can't the prefects take the classes so as to keep the whole school busy and working?' That, I was firmly told, would be an abuse of unqualified child labour. They would not know how to teach and they would be incapable of keeping the class in order. I still cannot understand what has got into the minds of our 'educational professionals' when they talk such utter nonsense. Put into positions of responsibility, most boys – especially those selected for the right qualities – rise to the occasion and benefit by the experience. If boys are not taught to accept responsibility whilst at school, when will they ever learn this important skill?

When, in due season, I became a school prefect, I admit I tended to be a little lenient. I gave out more 'verbal and written warnings' than others and had fewer entries in my 'Prefects Book'. When boys who persistently broke the rules and abused my leniency were eventually summoned to the Prefects' Court, I saw to it that the majority were given essays or lines to write. In two years as a school prefect I never caned a boy myself, not because I thought the practice barbaric, but because I did not have sufficient confidence to administer corporal punishment in a way which had the right blend of severity and leniency – a difficult balance to strike, or so it seemed to me.

Although I was at school throughout the War, when many of the younger masters were in the armed services, I was extremely well taught. The main talent, I now realise, was reserved for the specialist teaching in the higher forms, for the school's reputation depended greatly on the number of admissions it achieved into University, especially into Oxford and Cambridge. The masters who taught the 'Open Scholarship classes' were, almost without exception, First Class Honours Graduates from very good Universities and most of them had been Exhibitioners. Many of them remained active in their academic fields, continuing research, writing papers and books and participating in the work of their specialist learned societies. My main mentor was the Chief Science Master, Mr Cecil Prime, known to the boys as 'Cheese' (Prime cheese!) and he was responsible for opening up my mind to the fascination of taxonomy, plant physiology, ecology and genetics (see chapter 18). He used to lead field excursions on Wednesday afternoons and he expected all those boys who were 'interested' in his subject to attend. This created difficulties. Wednesday afternoons were reserved for official school games and also for the parades of the OTC, so one had to make invidious choices between attending Mr Prime's excursions, attending parades or playing games for the school. I used to achieve some sort of a compromise and when I became an NCO in the OTC, I was able to delegate some of my duties to others and so had free time to attend some of the field classes. Those who were good at games had more difficult choices. If they were playing for the school teams, they could not 'delegate', and thus they were only able to attend a very small number of these Wednesday afternoon excursions. However, there was no way of reaching a compromise with Mr Prime; you were either 'interested' and attended or you were 'uninterested' and did not. Due to masters like him, the Sixth Form environment at Whitgift was more akin to that of a university than to a school, and a school in war-time at that. I am extremely fortunate to have been taught by people of this high standard.

But there was one character whom both I and the school could have done without. He was employed by the OTC as the 'professional' soldier who assisted the part-time officers (serving members of the teaching staff) in the day-to-day running of the Corps. These 'professionals' tended to come and go on secondment from the army, as they were transferred from one posting to another. Towards the end of my school days a somewhat shady character was attached to the OTC as Regimental Sergeant Major (RSM). I was then a Colour Sergeant and for some reason or other, this new RSM took an instant dislike to me. I think this was due to the fact that he had a habit of coming on to the parade ground and ordering the NCOs to 'excuse' some of the cadets under their command for 'special duties', which normally turned out to be running private errands for himself. The more I objected to this, the more demands he made and eventually I was so frustrated with this undermining of my authority that I did an unheard of thing – I put him on a charge! Now it is quite improper for a mere Colour Sergeant to put a more senior ranking (and adult) RSM on a charge, but nevertheless that is what I did. I have always tended to be a hoarder and hate to throw anything away (much to my wife's disgust) and I still have the letter which I wrote to my Commanding Officer, which, inter alia, reads:

'I hereby accuse the above of:-

1. Reprimanding a senior NCO and severely criticising him in his absence, but in the presence of junior cadets in a manner prejudicial to good order and discipline.
2. Favouritism of certain cadets, especially by making a statement tending to cause resentment and ill-feeling amongst other cadets.
3. Knowingly and purposely giving misleading information in order to prevent a Certificate 'A' candidate from attending parade, in order to secure cadet labour for other purposes'

I ended my letter with the words, 'It is my wish, Sir, to have a positive assurance that the alleged charges are rectified in the future. If the accusations made, which I consider to be serious, are not proven, it is my purpose to tender my resignation and, with regret, to give up my rank and status in the OTC. I would add that I present this case not for my own advantage or satisfaction, but in the interests of the Whitgift OTC, in which I have had the honour and privilege of serving since October, 1941'.

I was duly summoned to attend a private meeting of the CO and his fellow officers. I was politely but severely told that I was quite out of order and that a mere C/Sgt. could not put a RSM on a charge. However, that being understood, I was told the points I made would be carefully looked into, but not by way of the 'charge' which I had quite improperly made. I was informed that there was no need for me to resign and that there were plans afoot to promote me to CQMS if I behaved myself in future. A year later, when I returned to the school on vacation from University, I learnt that the RSM had left under a cloud. I felt that my 'charge', although perhaps ill-conceived and improper, had not been entirely without reason. It may even have alerted the officers to the fact that something fishy was afoot.

The school offered a wide range of choices to develop the talents of the boys. There was an excellent school library, run almost entirely by the boys with a little help from the staff. It had a modest budget to purchase new library books and for a time I served as a school librarian and chairman of the school book selection committee. We took our duties very seriously, perusing many papers and magazines for reviews on new books and deciding which

6. Whitgift School Departmental Librarians, 1946. Author, seated, far left.

we could afford to buy from our small budget. There was also a school choir
and orchestra and each year impressive concerts were arranged, to which
parents and friends were invited. I never mastered an instrument (something I
have regretted ever since) but I did sing in the school choir until my voice
broke and I much enjoyed the team work which a large choir generates.
Although not very musically inclined, I also enjoyed the music, which, I was
told by those who knew about these things, was of a very high standard.

The Junior School, catering for boys between the ages of 9 and 13, had an
active stamp club and Hornby model railway club and I joined both for a
time. (Hornby model railways and Meccano sets were in those days necessary
and cherished possessions of most schoolboys.) All these activities, even in
the junior school, were largely run by the boys themselves, with a master
being 'allocated' to each activity to ensure that things were done properly
and that not too much time was taken up with these out-of-school-hours
events. As I look back, it is amazing that such a full and varied programme
was run at the school during the War years. When one considers that almost
every activity, from OTC parades to choir practice, was constantly being
disrupted by the wail of the air-raid sirens, it is a wonder that life continued
relatively normally during such very abnormal times.

Because of the frequent air raids, a large part of the parade ground was
dug up and underground air-raid shelters were constructed underneath.
When the sirens went, we had to march out of the school 'at the double' and

go into our allotted shelter. Lights were fitted, so in theory studies could continue even whilst the air raid was going on above us. Occasionally, however, the lights fused and the shelter would be plunged into darkness. This was an opportunity not to be missed, and we quickly evolved the idea of 'raiding' the shelter next door, banging as many boys as we could find on the head with our text books, then beating a hurried retreat back to our own sanctuary. Unfortunately, with no lights to assist us, it was impossible to distinguish boys from masters and on one of these forays I misguidedly bashed a somewhat prickly master on the head with a heavy text book. At that moment the lights came back on and I was discovered, together with two or three of my companions, with arms raised in threatening fashion, armed with a suitably weighty text. The master was not amused and the three of us were summoned to appear outside his office as soon as the air raid was over. We were all given three strokes of the cane and although I appreciated they were well deserved, I thought it very unlucky that only three of us were caught in the act whereas the whole form had been involved in the foray. Life, I then learnt, is not fair and it gets less fair as time goes by. This is an important lesson and all my children knew what answer they would be given when they complained to their mother or to me about the inequalities of human existence.

School was exciting and pleasurable and, when I eventually had to leave to go up to University, I was only too glad to be allowed to return for an extra 'unofficial' month, since the school term started four weeks before the University year commenced. For this brief month I was promoted to the giddy heights of Company Sergeant Major. I didn't have any classes to attend or laboratory practicals to perform. I had the run of the school and was still allowed to retain my status as a senior prefect. In fact, life was great and I wasn't quite sure I wanted to move on to the next, unknown, stage of my career.

For many years after leaving school, I went back for a half-day each vacation to look up some of the masters who had done so much to help me in my studies and to inculcate a spirit of inquisitiveness and self-confidence. On several occasions after my return to the UK, I have planned reunions for the Old Boys of the school living in my area and it is clear that the affection I feel for my *alma mater* is shared by many other boys, both senior and junior to me. I sense that this 'Old Boy spirit' is not nearly so apparent in many schools today, where the common attitude of the pupils on leaving seems to be: 'Thank goodness my schooldays are over!' In my case, however, I look back on those school years with nostalgia tinged with a modicum of pride.

It is impossible to state exactly what responsibility Whitgift School can

7. *The author with senior members of the Officers Training Corps, 1946.*
Battery Sergeant Major WOI D.C. Edwards, seated fourth from left.
The author, Colour Sergeant, seated third from right.

claim for shaping my subsequent career, but I am sure in my own mind that I owe a great deal to the masters, especially the science masters, the officers in the OTC and also to the leading boys in my house, in the Science Sixth and in the Prefects' Room. The influence of a good school gave me self-confidence and helped me to make decisions without too much difficulty. Once one had made up one's mind on a course of action, one had to live with the consequences, whatever they might be. It taught me the importance of working together as members of a team, for by so doing one could achieve a great deal more than by acting independently as a lone individual. It was abundantly clear to me, when I left school, that privileges were intimately tied to responsibilities. As one of my OTC superiors, an ex-cavalry officer, was always telling us: 'To be a good cavalry officer you must put your horses first, your men second and yourself last!'

CHAPTER 3

I Choose Farming

'Whoever could make two ears of corn or two blades of grass
grow upon a spot of ground where only one grew before,
deserves better of mankind, and does more essential service to his community,
than the whole race of politicians put together'

Jonathan Swift

I CANNOT PIN DOWN exactly when I wanted to 'become a farmer', but perhaps it was a year or two before the War, when my parents decided to spend their holiday in Devon and chose to stay at a farm-house rather than an hotel. My father was intent on a 'touring' holiday. He had just bought his first second-hand car and was anxious to try it out on the small winding lanes of the south Devon countryside. They booked a fortnight at Shadrack Farm near Totnes. It is still there, pretty well unchanged since the late 1930s, and there it was that I decided to 'opt out' of the 'driving around Devon' holiday and spend my time 'working' on the farm. My father was not a particularly good driver, having only learnt to drive when in his fifties, so I was happy when given permission to stay behind, helping with the harvest.

The farmer and his wife were marvellous. Having me as an extra unpaid hand – albeit small and inexperienced – was no trouble to them and I spent the first long, warm summer day helping to stook sheaves of corn as they were dropped behind the reaper-and-binder. This was hard work; the corn was dry and full of thistles and the sheaves so thick it was difficult to put my arms round two at once and bunch them together to form the stook. But I persevered and I enjoyed the adult company of the Devon farm workers, who were kindly and indulgent in teaching me my first lessons in 'bringing home the harvest'.

The day was over at about 6 p.m., and at 7 all the family and their guests assembled in the large farmhouse dining room for their 'high tea'. There were about a dozen of us round the solid oak table and, this being the first evening, my mother tried to break the ice by asking me what I had been doing all day. 'Mummy', I said, 'I've had a wonderful time. I've helped pick up the b—s and put them into f—g heaps!' You could have heard a pin drop. My mother went puce and ordered me to bed. As I left the room, I

8. The author feeding chickens at Shadrack Farm, South Devon, 1938.

could hear her apologising for my behaviour and I remember the smile and twinkle on the faces of some of the men. I went to bed hungry, wondering what on earth I had done wrong. I thought I had earned a little praise for my hard work but all I had got was a humiliating order to an early bed. Somehow, however, this did not put me off farming, nor did it prevent me getting up early the next morning to help the farmer's wife look for newly-laid clutches of eggs hidden in the hedgerows and loose straw around the farm yard.

About a year later, I asked my father to allow me to spend a whole summer holiday working on a farm. The War had already started, so it was thought desirable that I went to a farm to the north-west of London, in order to escape the attentions of the Luftwaffe, which concentrated on the south-east. In pre-war days, 'farm pupils' paid farmers to teach them the rudiments of agriculture, receiving only their keep during their 'training year'. My father found what was deemed to be an appropriate farm in the neighbourhood of Northampton, and I was duly taken there by car for the start of what should have been a rewarding working holiday over the harvest period.

The farm looked pleasant enough and we were greeted by a kindly woman who explained that the farmer was 'at the market' but was expecting me. Off-loading my little suitcase from the car, I was shown into a small bed-room at the top of the large farm house. My parents duly departed and left me to my fate. A couple of hours later the farmer arrived. He was a tall, gruff person with a very red face. He seemed a little unsteady on his feet and was abominably rude to the kindly woman, who turned out to be his housekeeper. I was told to be up early next morning in order to help him harness his horse to the farm cart, as we had to 'go to market' for some important business. Breakfast was a strange affair with little to eat but plenty of hot coffee to drink. Then we were out into the yard preparing the horse for his trip. The day was sunny and bright and I was ready and prepared to learn some basic principles of farm husbandry. The farmer seemed less gruff and more jovial, but just as red in the face.

We arrived at the market and I was taken round some of the stands and introduced to some of the farmer's companions, who seemed amused to find me in tow with nothing much to do and obviously very green. After a couple of hours spent doing nothing in particular, the farmer said that he had some important private business and I was to wait with the horse and cart for his return which 'would be in just a few minutes'. Some three or so hours later the farmer returned, somewhat unstable on his feet, and ordered me to drive him home. This was the first time I had handled a horse, let alone a horse with a cart attached, but I managed to remember the long way back to the farm. On arrival, instructions were bellowed for 'food for two hungry workers' and half an hour later the housekeeper produced an excellent late midday meal, which I thoroughly enjoyed, having had little to eat at breakfast, whilst my farmer-tutor drank a pint or two of beer and promptly fell asleep.

A couple of hours later I was told that the farmer had another important visit to make with the horse and cart. I had at least learnt how to harness the horse and hitch him to the cart. After another ride of about six miles we came to a village, where I was once again ordered to wait with the horse and cart while the farmer carried out his urgent business. The day wore on and evening arrived, with still no sign of the farmer. Thinking that I had misunderstood, and that he had meant me to drive the horse and cart home, I wended my way back to the farm on my own. The housekeeper seemed worried to see me and even more concerned when I related my experience. She obviously knew what was in store for me, for several hours later a very irate, scarlet-faced farmer roared up the farm drive, demanding to know where I had been all evening and calling me a large number of strange names which were all new to me. The housekeeper was very concerned about the whole situation, and tried to explain to me that he was always like this 'when in his cups' but that I shouldn't take it personally or worry unduly about getting things wrong. In spite of her reassurances, I got the distinct impression I wasn't wanted and when eventually the farmer fell asleep in a

9. Threshing corn in the winter holidays at Fawkham Green, Kent, 1942.

chair in the kitchen, I plucked up courage to borrow ten shillings from the housekeeper, packed my suitcase and walked a mile or two to the nearest main road to hitch-hike my way to the nearest railway station so as to catch a train to London and thence to Sanderstead. My parents did not seem too pleased to see me back so soon and were even less pleased when they received a phone call from the farmer complaining that he had been deceived by his pupil, who had disobeyed his orders and who, in his opinion, should be horse-whipped. I escaped the horse-whipping but was told very clearly that any other 'farming holidays' were out of the question and that in future my holidays were to be of a more conventional kind.

However one of my twin brothers – Donald – heard of the affair. He was then serving in the RAF at Fawkham Green in Kent and he was billeted on a small holding run by a Mrs Self, a farmer's widow, who provided bed and breakfast. Donald asked if I could be allowed to work on the small holding in return for my keep and this was duly agreed, with the proviso that I shared a small attic bedroom with the farm worker – Ken Smith. A few days later I cycled from Sanderstead to Fawkham Green and spent the remainder of my holiday very happily at Woodlands Farm. I enjoyed the experience so much that I spent the following two summers working as an unpaid farm worker, learning a very great deal from my goodly mentor, Ken Smith, and making friends with my brother's four RAF colleagues, who were also billeted on Mrs Self's farm. I pay tribute to Ken Smith in a later chapter. He was assiduous in teaching me new and exciting practical skills, some of which I remember and practice to this very day.

As my days at Woodlands Farm were confined to the Easter and summer holidays, I looked for opportunities to extend my farming experience at weekends during term time. There were no farms near to Sanderstead, as 'suburban creep' had covered most of the countryside with new housing estates and the areas not developed in this way were wooded or designated as 'bird sanctuaries'. However, some ten or so miles from home suburbia gave way to open fields and, on one of my numerous cycle rides into these unspoilt areas, I stopped to mend a puncture by the side of some farm workers' cottages. Needing some water to locate the puncture, I knocked at the cottage door and explained my problem. A gorgeous girl in Womens' Land Army uniform attended to my needs. I later learnt that she was named Beatrice. A friendship developed and my cycle rides wended their way more and more frequently to Beatrice's cottage and thence to the farm on which she was employed, so I rapidly developed a second string to my farming practice, helping Beatrice and the other farm staff at whatever tasks they were allotted.

I learned that Beatrice had a birthday coming up, and mentioned this casually to my father. He was a perceptive and kindly person and, realising that Beatrice was rather special to me, suggested I should buy her a small present Knowing this was more than I could afford, he gave me the money to buy a magnificent box of chocolates which, at the appropriate time, I duly presented to her. In thanking me, and realising that a degree of puppy love was developing, she gently informed me that she was 26 years old and engaged to be married! I was crestfallen and very embarrassed, but expressed the hope that this would not prevent me from seeing her and assisting her with farm chores whenever I could manage the time to cycle over. I must have been about 12 at the time.

My interest and love for farming continued through my adolescent years, although other commitments – not to mention my studies – obviously made major inroads on my time. When I was about 15, I was beginning to make my mark with Mr Cecil Prime, my school botany master. He obviously had me marked down as a potential biologist and went out of his way to interest me in various extra-curricular activities. He was then running a Natural History Society in Croydon and invited me to several of their evening

10. Harvesting by reaper and binder at Fawkham Green, Summer, 1943.

meetings. He was also, as part of his War effort, co-operating with Rothamsted Experimental Station in undertaking some agronomic field trials on a section of the school grounds. He invited me to assist with planting the experimental crops and applying a range of fertilisers and pesticides. I mentioned that I was very interested in the actual field work, which he found very strange. When he found I had distinct leanings towards an agricultural career, he was quite put out, saying that agriculture was merely an exercise in man-made ecology and that studying the 'real thing' in the wild was much more intellectually exciting. He never became reconciled to my wish to study Agriculture rather than Botany at University and perhaps felt that I had 'missed my vocation'.

This perception of a 'missed vocation' was shared by my father, who from an early age had marked me down to become an accountant. I was proficient at maths and my father thought that this meant becoming a chartered accountant and earning a good living. There was no money in the family so any thought of 'farming', which presupposed the expenditure of large sums of money on the acquisition of a fully equipped farm, was completely out of the question. My father was prepared to find the wherewithal to pay for me to be 'articled' to a London firm of chartered accountants, but was not prepared to help me in any way if I rejected this offer and opted to 'dabble' in agriculture. He made it clear that 'If farming is what you want, go to it but you are on your own!'.

In my sixth form at school I had successfully achieved all the necessary qualifications to go to University to read agriculture and was determined to do two things. Firstly, to endeavour to avoid what was then a compulsory pre-entry 'practical' year on a farm and secondly to gain exemption from the first year of the course, which consisted of botany, zoology, chemistry, physics and geology, all subjects which I had taken at school, passing the necessary examinations to prove it. But getting to University in 1946 was itself a problem. 1946 was the peak year of entry for ex-service men and women and the number of entrants taken direct from school was minimal. My choice was restricted to Wye College (London University), Reading and Nottingham and my first choice was Wye. It had all the attributes I thought desirable. It was a small College with a long academic tradition, having been founded by Cardinal Archbishop Kemp in 1447. It awarded London degrees and it was situated in first-class farming country – fruit and hops, arable crops, cattle and sheep (177).

I was determined to get to Wye, so I applied for, and was granted, an interview with the Principal, Dunstan Skilbeck, and his Vice Principal, Norah Penston. My mode of transport to Wye from Sanderstead was by bicycle, a journey taking about three hours, and I was a little embarrassed that my riding kit was not really up to the standard of dress required of an interviewee. Skilbeck was a stern upright man with an ability to fix his eye intently on the candidate in a manner calculated to undermine his self-confidence and fill him with awe. Norah Penston was a prim, somewhat Victorian-looking lady who did her best to emulate Skilbeck's stern countenance, without quite succeeding. The interview was tough in every respect. It took place in a large formal room and the dark wood panelling added a sombre and awesome appearance to the chosen setting. It was quite clear that Skilbeck, an ex-Wing Commander in the RAF with a distinguished War record, was not at all keen to take in schoolboys 'wet behind the ears'.

I was given a long and exacting oral grilling, during which I grew more and more despondent. Eventually Skilbeck, magnificently robed in his Oxford cap and gown, leaned across the table to me and said, 'Well, Wilson, if we take you in, which we probably won't, what do you think you will be doing in ten years' time?' This is, of course, a common question to ask at interviews, although usually at a much later stage than pre-entry to an undergraduate course, and I hadn't a clue as to how to answer. After a little hesitation I replied, 'Well, Sir, I really haven't got round to thinking about that question but, since you ask me, I rather think I might like your job!' Luckily Skilbeck had a very good sense of humour. After giving me a prolonged quizzical look, he smiled and said 'Well, Wilson, I am probably about to make a terrible mistake, but you're in!' Norah Penstone did not smile. She regarded me with great suspicion and appeared to take my innocent and honest answer to a difficult question to indicate insubordination verging on insolence.

Thus it was that I proceeded to Wye College on the next (or was it really the first?) major step of my agricultural career. It transpired that there were only six students admitted straight from school – two boys and four girls. All the rest of the undergraduates were ex-servicemen and women, justly proud of their active service record during the War and somewhat suspicious of 'mere kids from school'.

CHAPTER 4

War Memories

'Now tell us all about the war,
And what they fought each other for.
But what they fought each other for
I could not well make out'.

Robert Southey

I WAS ELEVEN YEARS OLD when War broke out. I was utterly convinced we were fighting a 'Holy War' and that in the end 'Right' would prevail. Nazi Germany and Adolph Hitler were beyond the pale and must be exterminated at all costs – the sooner the better. Although my grandmother had rightly guessed that the War would be protracted and bloody, it seemed to me in 1939 that it would all be over very soon. After all, the Maginot Line in France was impregnable and I was brought up to believe we had the best army, navy and air force in the world. With the backing of our Commonwealth friends and other allies, who could prevail against us?

So I took the usual pre-war precautions – the digging of Anderson air-raid shelters in back gardens, the issue of gas masks and ration books to everybody and the printing of Identity Cards – with a pinch of salt. The gas masks were taken very seriously indeed. Everyone, from the very young to the exceedingly old, had to carry these masks with them at all times, safely tucked into a small square brown cardboard box. Some enterprising suppliers manufactured ornate covers to go over the brown boxes, so that they looked more elegant as women carried them faithfully to dinner parties or dances. It was as well to be prepared but surely, we thought, all these measures were not really necessary. German bombers would never reach the London area, food would still be imported from America and the Commonwealth and life would soon return to normal.

One of my twin brothers, Arthur, was in the Palestine Police during the first part of the War but the other twin, Donald, joined the Special Constabulary before the outbreak of hostilities and left home regularly in his brand new uniform for his policeman's duties, full of pride and importance. Both my father and mother joined the Air Raid Precaution (ARP) and became wardens. This meant my father giving up three or four evenings a

31

11. Air Raid Precaution Wardens, 64 Group, Sanderstead, 1940. Father, seated, third from left. Mother, standing first row, fourth from left.

week – after his return from a day's work in London – to man the air-raid post, conveniently set up in a garage adjacent to our home. My mother undertook day-time warden's duties, five or six half-days a week, and I volunteered my unofficial services as a 'bicycle runner', able and willing to travel quickly between any local 'incident' and the headquarters of combined operations sited in the nearby air-raid post.

My father went further, as he was a very thorough person who always wanted to do things properly and to the best of his ability. Although we had no air-raid shelter in our own home (we agreed with our neighbours to share one next door), we converted our drawing room into a gas-proof room by sealing the windows with sticky tape and making a second door to act as a safe entry- and exit-point. We exercised diligently in putting our gas masks on and off and my father insisted that we sometimes wore them whilst 'working' – doing household chores or gardening – so that we grew familiar with the style of breathing necessary when wearing the awkward and uncomfortable device.

In these days a call went out for scrap iron. The story was that such materials were necessary for the war effort, as they could be melted down and used to make guns and ammunition. My father suggested that I might perform a useful and patriotic function by visiting every house in our road with a wheelbarrow and asking if I might take down its owner's garden railings and chain fences and cart them off 'for the war effort'. A surprisingly large number of good folk cheerfully agreed, and after a few days I had accumulated a large pile of chains, railings and other metallic objects ready and waiting to assist the war effort. It transpired that such rusty old iron, covered in a variety of paints, was unsuitable for the manufacture of guns and ammunition and was never used. I was amused when, shortly after the War, I was approached by an elderly couple who lived along our road. 'When can we have our railings back?' they asked. I never learnt the eventual fate of all this scrap iron, but still think that the person responsible for running this campaign, which was nationwide, should have done his homework a little more carefully before the massive collecting operation got underway.

I clearly remember Neville Chamberlain's dramatic announcement on the radio: 'As no reply has been received (from Hitler), a State of War therefore exists between us'. This, I thought, is it! My concern was reinforced when, shortly afterwards, the first 'real' air raid siren wailed and we thought we were in danger of imminent attack. But it was a false alarm and as day followed day and nothing much happened, I thought that the concept of a brief War must be correct. When the radio announced that 'Gerry' had dropped a couple of bombs on the Isle of Wight, killing a few rabbits in the process, I thought that the War was indeed 'phoney' and all our detailed precautions a complete waste of time. My confidence was badly shaken a few weeks later. I was playing in the garden with some friends and we noticed a large formation of aircraft flying high overhead. We all pretended to identify the planes, as aircraft recognition was then a popular children's game, but although we couldn't settle on the right name, we were all agreed that the aircraft were 'friendly' – possibly a squadron of newly-arrived aircraft at Biggin Hill on a training flight. Imagine our surprise when the planes started hurtling downwards and dive-bombing Croydon aerodrome, about three miles away. We rushed into the nearest air raid shelter with great alacrity, a little shaken and ashamed of our lack of prowess in recognising enemy aircraft when we saw them.

A turning point in my youthful appreciation of the fortunes of war came when Churchill came back to power as Prime Minister. I was too young to have had any recollection of his earlier days in the forefront of British politics, but it was quite clear to me that here was a man of vision, a natural leader who could command our complete confidence and trust. I was amused when I read that the small West Indian island of Barbados had sent a telegram soon after his appointment was ratified by the House of Commons: 'Carry on, Britain. Barbados is behind you!' I also chuckled at the message which was flashed round the British fleet, 'Churchill's back!'. Our faith in Winston Churchill was fully justified and I still regard this war-time Prime Minister as one of the nation's greatest leaders. After the War I felt badly when Churchill was rejected by the electorate, who favoured Attlee as the first post-war Prime Minister. At the same time I realised that when he made a return to the forefront of British politics a few years later, he was past his prime and his dominant position as the nation's obvious leader had been lost. When I watched his funeral service many years later, and listened to Richard Dimbleby's moving commentary, I shed a real tear.

The event which made the greatest impact on me during the early War years was the evacuation from Dunkirk. Our rapid retreat from the Low Countries and northern France was full of doom and gloom and when what was left of the British Expeditionary Force (BEF) had their backs against the sand dunes in Dunkirk, it seemed that all was lost. We reverted to what seemed to me futile defence policies with an imminent invasion in sight, such as turning round finger posts on country lanes to 'confuse the enemy as they advanced', and the rapid building of countless thousands of concrete 'anti-tank fortifications' all over the country. It seemed foolish to believe that a small pile of concrete some six feet high could prevent a determined armoured advance preceded by heavy artillery bombardment or a well-directed bomb, which would quickly have reduced one of these concrete blocks to rubble. Although it was not acceptable to think or speak defeatist thoughts, it was clear that everyone was expecting the worst. The Germans would throw the BEF into the sea and a few days later would invade the south of England in force, travelling as rapidly across southern England as they had done through Holland and Belgium. But, as all the world now knows, that was not to be. 650 small boats, some of them no bigger than in-shore pleasure craft, fished the brave soldiers of the BEF and their allies out of the sea in spite of ferocious air bombardment by the Luftwaffe. If ever victory, of a sort, was plucked from the jaws of total defeat, this was it. Most of our troops were safely back on British soil, and the expected invasion across the Channel never came. Throughout this most awesome time, Churchill was wonderful; he continued to make stirring speeches to parliament, to the nation and to the returning troops and I especially remember his riposte to Hitler, who had boasted that he would wring Britain's neck like a chicken's: Winston replied 'Some chicken, some neck!'.

So we had a temporary reprieve, but the War dragged on, going at times from bad to worse. Had it not been for the optimism and inspiration of

Winston Churchill, our morale would have sunk many times to a very low ebb. The 'Battle of Britain' was fought in the skies over and above Sanderstead and we watched with bated breath to see who would shoot down whom. One of our pastimes was to walk round the streets after a dog-fight had taken place, collecting pieces of shrapnel, which fell out of the sky like hailstones after each encounter. Sometimes we were lucky enough to find a spent machine gun bullet and on one memorable occasion, whilst taking part in a cycle ride into the nearby North Downs, we found an entire machine gun. It had obviously fallen out of a plane which had exploded in the air. We buried the machine gun in a wood, carefully marking the spot before going off to report it to the nearest police station. Imagine our embarrassment when we returned with a constable to the marked site, only to find that the machine gun had mysteriously vanished! Were there German spies about who had retrieved the gun, or had someone else stumbled over our find and taken it away as a war souvenir? We never knew, but the policeman regarded us with a very jaundiced eye and reprimanded us for 'wasting police time'.

The country to the Southeast of London was 'zoned' for the purpose of protecting the capital. Nearest to the coast were 'open skies' to enable the fighters to engage the bomber formations. About ten miles out of London was the second line of defence, a ring of barrage balloons, the object being to bring the bombers down by entrapping the planes in their wires as they descended to a low level in order to bomb London. Sanderstead was at the junction of these two zones. To the SE of us we watched the dogfights and to the NW of us we placed bets on whether or not the enemy planes were flying low enough to be brought down by the barrage balloons. Whichever defence was successful, the German planes crashed in and around Sanderstead, so in a sense many of our 'incidents' were self-inflicted, resulting from what in other circumstances might be called 'friendly fire'.

In 1940, after Dunkirk, the Germans paid us the compliment of dropping down 'oil bombs', which they had apparently collected from the ammunition dumps left behind by British forces in France. The object of the oil bombs was to cover about one quarter of an acre with thick oil, which was then ignited by a series of incendiary devices dropped with the bomb. One such object fell in our garden, a few yards from the back of the house. I was sleeping downstairs at the time, using the converted gas-proofed drawing room as a make-shift bedroom, this being regarded as a 'safe place' for me at night should the expected gassing campaign commence. I woke up to a terrific bang, and when I put on the light, I found a large piece of shrapnel from the casing of the oil bomb smouldering on my pillow an inch or two

from where my head had been lying. The garden was an awful mess –
covered in oil and with a large crater just beside the boundary hedge.
Thankfully, the oil did not ignite as the incendiary devices failed to work,
otherwise the house would certainly have been ablaze. I asked to be allowed
to sleep in my own bedroom from then on. I turned this incident to good
advantage by charging friends and passers-by 6d. to 'come and see the oil
bomb crater', which they did in quite large numbers. I made over ten pounds
and donated it to Mrs Churchill's fund to assist the Russians, who then had
their backs to the wall as the Germans advanced almost up to the gates of
Moscow. I received a charming personal letter back from Mrs Churchill,
thanking me for my generous gift and saying that this was the first time that
she had heard of anyone raising money by taking advantage of the blitz.

After the terrible night-time strafing of London, when it seemed all
London was ablaze, came the Doodle Bugs. We were still surrounded by
barrage balloons, floating on top of the North Downs in an attempt, not
only to bring down enemy bombers, but also to intercept the Doodle Bugs
before they reached their London destination. Yet again the suburbs, it
seemed, were expendable and it didn't matter too much if planes or doodle
bugs dropped out of the sky in and around Sanderstead. The result was that
the area to the south and east of Croydon became the resting place for
literally hundreds of bombers and Doodle Bugs. A map of the area produced
just after the War looked like a pin cushion, with each dot representing a
crater made by bombs, planes or Doodle Bugs. It seemed incredible that
anyone could survive, yet the number of fatalities was relatively small,
although many folk were injured, some very severely. The chief cause of
concern was the loss of property. About half a dozen houses in our road of
some 100 properties were either destroyed or so badly damaged that they
had to be pulled down and rebuilt. In our small stretch of road, however,
only the oil bomb fell out of the skies into our garden and thankfully the
ARP HQ in our neighbour's garage carried on its good work of defence co-
ordination throughout the duration of the War.

My father was a good administrator and his talents were soon recognised
by those in charge of the combined services of police, fire and ARP. He was
promoted from Air Raid Warden to an 'Incident Officer' which meant that,
on returning from his day's work in London, he could be ordered to go by
car anywhere in his 'District' (a few square miles around Croydon) to take
charge of any 'major incidents'. A major incident was defined as one
requiring a co-ordinated effort of all the emergency services and he had to give
the appropriate orders to the liaison officers of each individual service. This
was a tricky assignment, requiring great diplomacy, for Police Super-

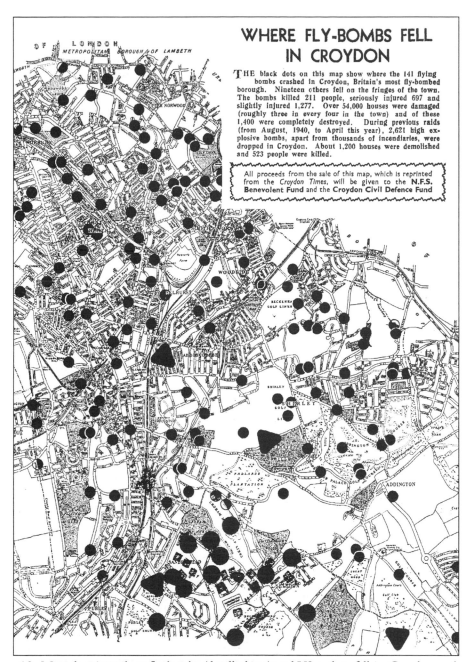

WHERE FLY-BOMBS FELL
IN CROYDON

THE black dots on this map show where the 141 flying bombs crashed in Croydon, Britain's most fly-bombed borough. Nineteen others fell on the fringes of the town. The bombs killed 211 people, seriously injured 697 and slightly injured 1,277. Over 54,000 houses were damaged (roughly three in every four in the town) and of these 1,400 were completely destroyed. During previous raids (from August, 1940, to April this year), 2,621 high explosive bombs, apart from thousands of incendiaries, were dropped in Croydon. About 1,200 houses were demolished and 523 people were killed.

All proceeds from the sale of this map, which is reprinted from the *Croydon Times*, will be given to the **N.F.S. Benevolent Fund** and the **Croydon Civil Defence Fund**

12. Map showing where fly-bombs (doodle-bugs) and V2 rockets fell in Croydon and District during the war. Circles, fly-bombs (Small circles, Croydon area; large circles, Sanderstead area). Triangles, V2 rockets.

37

intendents and Chief Fire Officers did not take too kindly to obeying the commands of a 'part-time civvy'. However, he carried out these duties without too much criticism and, as the Blitz wore on, it must have been obvious to everybody that some such co-ordinating role was essential.

If an 'incident' was near to home and not too late at night, I used to assist my father by acting as an 'incident runner' on my bicycle. This was quite an important duty, as major incidents often resulted in broken telephone lines and blocked roads, so that conventional means of communication were unusable. The army used mobile radio sets in battle; planes and ships communicated by radio but, for some technical reasons which were never made public, it took until the end of the War before the civilian emergency services had the routine benefit of the use of the air waves.

After the War both my mother and father were awarded Civil Defence Medals for their voluntary work in the ARP and justly so. I was a little miffed that I was too young either to be recognised as an official member of the ARP or to get a defence medal. I did not even have an arm band to indicate my position, although by then I had risen through the ranks in the Officers Training Corps at school. Military uniform, however, was not allowed to be worn when not on strictly military duties! When not on parade in the OTC, I was just a simple civilian schoolboy.

When houses were badly damaged by the Blitz, families had to be evacuated and, wherever possible, attempts were made to salvage their goods and chattels and to take them to a central storage facility 'somewhere in the country'. The whereabouts of these war-time depots for personal effects were never made known, nor do I know how efficient the system was in returning the goods to their rightful owners at the end of the War, or when the people affected had been re-housed. This was important and arduous work, always requiring more pairs of hands to shift and box the furniture than were available. So, when not acting as an ARP runner, I used to do my bit at moving furniture from bombed properties on to lorries, vans or pick-up trucks and I developed quite a skill at efficiently boxing up glassware and crockery. This skill came in useful later, when we had to pack our own goods and move house from one African posting to another.

After Doodle Bugs came the V2 rockets. These were considerably more powerful and destructive than the lumbering old Doodle Bugs and were capable of destroying not just a house or two, but a whole street. Travelling, as they did, more quickly than they speed of sound, they came without warning; their 'whine' was heard only after they had landed. With the Doodle Bugs, there was a period of about half a minute between the cutting out of their engines and their descent to the ground, so that one had time

13. Local Fly-bomb (Doodle-bug) damage, 1943.

to take cover. Indeed, some of us got quite clever at predicting where a Doodle Bug would land, having noted its direction or flight, its height, and the time the engine cut out. With practice, one could guess correctly the bomb site to within a street or so and take any evasive action that might be required.

But the 'real War' was being fought in foreign lands by distant armies and by the navy in far-away oceans. One felt instinctively that it was in such places that the War would be lost or won since, although the Blitz was terrible and the death and destruction wreaked was immense, it seemed obvious that bombing alone – however prolonged – would not win or lose the War. This was true and it seems strange that, in more recent times, this lesson does not seem to have been learnt by our modern war lords. Bombing Vietnam, Iraq and – more recently – Serbia and Chechnia does not defeat a nation; indeed it tends to make it more resolute in its resolve to continue the fight. It is only land battles, with the army aided and abetted by the navy and with support from the air, that win or lose wars, as the Second World War was to demonstrate so dramatically. So it was the land battles in Russia,

Africa, the Far East and later in Europe that one watched fearfully from afar, often with bated breath.

I had large-scale maps of North Africa, Europe and, later, the islands of the Pacific in my bedroom and every day I plotted the advance or retreat of the warring armies. The one field of battle which is still imprinted vividly on my memory was the North African campaign and the eventful duel between Rommel and Montgomery. I was impressed that Monty spent much time researching the mind and past battles of Field Marshal Rommel. I tried to remember the importance of 'knowing your enemy' when planning the tactical operations of our annual OTC field days, where the whole corps went by train to some appropriate open heath or common, dividing into two groups, the 'defenders' and the 'attackers'. I used to work out the probable strategy of my opponent and deploy my own men accordingly. It was quite remarkable how relatively easy it was to calculate whether he would be most likely to mount a full frontal attack, a flanking movement or a pincer movement to the rear. Different commanders favoured different tactics and it was by knowing who was in charge of the enemy forces that one could best estimate how to win the engagement. Similar calculations must go through the minds of good rugby players, who seem to have the happy knack of being in the right place at the right time. Alas, although tolerably good at this guessing game on the mock battle field, I could not bring the same skills to bear on the sports field. My opponents were too quick for me and clearly in the middle of a gruelling rugby game one had much less time to work things out in a careful logical manner.

My war-time exploits were not entirely tied up with the ARP and the OTC. In addition to studies at school, I was slowly working up my enthusiasm for matters agricultural and I was also a keen and active member of the Youth Fellowship (YF) attached to our local church in Selsdon. We met in the bell tower of the church, but only when it had been restored, since the east-end and chancel were badly damaged by a bomb and were out of action for a year. The tower was situated over the north transept and luckily escaped major damage, so our 'meeting place' was in weekly use for most of the War. The YF had a social as well as a serious side and we organised public dances, cycle rides, visits to places of interest in the nearby area and, in the school holidays, YHA tours. I led the YHA group for several years, and was followed by John Cannon, who was also a contemporary of mine at Whitgift School. We produced a monthly magazine which some of us laboriously typed out with the help of carbon papers (this was before the days of duplicating machines) and each year we presented a Nativity play which played to 'full houses' for several days before Christmas. A great deal of thought went into selecting a girl to play the part of Mary, and this went in turn to Jean Holdway, Betty Finch, Margaret Herbert and Barbara Osborne. Some of the more musically inclined, led by a very accomplished pianist and composer, Peter Jeffree, formed a musical appreciation group known by the novel title of *Allegro ma non troppo*. The YF provided the first chance to meet members of the other sex in a relaxed way, and many boy- and girl-friendships were made at that time. These were invariably 'innocent' relationships. Most of the dances were attended by a group of couples rather than just by one lad and his lass. I made several close girl friends in this way, very special among them being Barbara Osborne and Betty Finch, and although in my case marriage was not the eventual outcome, there were several instances of these early liaisons resulting in happy life-long unions. Thus my good friend John Sclanders married his YF girl Jean Holdway; John Cannon married Margaret Herbert and Derek Holdway married Evelyn Johnson. There could well have been other happy marriages which took place after I had left to go to University, but I do not recollect them. The YF provided a very helpful and healthy means of dealing with those difficult and explorative teenage years and I feel sad that it no longer seems possible for the young folk of today to have such a constructive and innocent outlet for their early sexual development.

So the War years, although fearful and stressful at times, were strangely enjoyable too. The War gave me opportunities and self-confidence for taking many initiatives and assuming responsibilities and, as I look back, it was a wonderful time in which to grow up. There was a great deal of team-work and 'pulling together', whether at school, in the OTC, in the ARP or in the YF. We were single-minded in our resolve to make the best of the War and to express ourselves to the full. I am indeed grateful to have had this experience and to have shared it with many good and close friends, many of whom are still in touch with us today.

But if you asked me at the time whether or not we really had to go to war and fight for six long and bloody years, drop two atomic bombs on Japan and kill more innocent civilians than in any previous global conflict, then, as Robert Southey remarked, 'I could not well make out'.

Chapter 5

Salad Days at Wye College

'Remind me to remind you, we said we'd never look back'
Julian Slade

BEING ACCEPTED INTO Wye College was not quite the same thing as getting there! There were fees to be paid and expenses to be met. On the strength of my showing in the Higher School Certificate examinations, I was granted a 'Major County Scholarship' by the Surrey County Council. The problem was that this 'major' scholarship was means-tested and I was informed that, because of my father's income, the scholarship would have no monetary value. My father was adamant that, whatever the fools at County Hall said, there was no way he could find the fees and my keep. I would have to think again and possibly reconsider my refusal to take up chartered accountancy. I was equally positive that I wanted to go to Wye College, so I hopped on my bike, cycled to Kingston-on-Thames and politely asked to see the Director of Education.

It seems quite incredible that, in 1946, a young school boy could arrive unannounced at County Hall and expect to be granted an interview with the Head of the Education Department. Nevertheless, after quite a long wait I was duly ushered in, wearing my cycling togs, to the Director's office. I told him I wanted to read agriculture and that I thought this was the most important subject one could study at University. I pointed out that we were still rationed in the UK and needed more food and that only a more progressive scientific agriculture could provide it. That was why I was determined to read for an agricultural degree.

I was politely heard and was asked why my father would not support me. I explained the reason. I was asked what my Head Science teacher at school had to say about me. I had taken the trouble to bring a reference with me, which I showed my interviewer. 'I see Mr Prime would prefer you to go to Cambridge to read botany', he said. 'Yes', I replied, 'but that would be even more out of the question, for the tuition and College boarding fees at Cambridge are much higher than at Wye College.' He was noncommittal, but I felt he had given me a fair hearing. If I failed, at least I had tried. I cycled home, wondering what my next step would be. I did not have long to wait. In the following morning's post was a letter to my father saying, 'That, taking account of new factors which had been brought to his attention, the Director of Education has agreed that my major scholarship could be valued at £230 per annum. Would my father please signify his acceptance?' My parents were still sceptical about the wisdom of studying agriculture, a subject which in their opinion led only to a career in practical farming, but as it was clear that my mind was made up, they graciously agreed and wished me well.

So I got to Wye. As I had hoped, because of my school examination passes, I was exempted from the first year of study and was entered into the second year of the 3-year course. Most of my fellow students in my class had been at Wye for a year already and were well settled in. There were a few ex-servicemen who had started their degrees at Wye before the War interrupted their studies and were now coming back for their final two years. There were no other ex-school students in my year, although in the parallel class studying horticulture there were five others straight from school, one boy and four girls.

Life was a little tough. The whole atmosphere of the College was dominated by ex-servicemen with all the camaraderie of the officers' mess to bind them together. Not only that, they had relatively generous ex-servicemen's' grants which were more than double my Surrey major scholarship, so they could afford to live well; certainly they did not have to count their pennies before going to the pub each night. My tales of days in the OTC, latterly as a CSM, cut no ice whatsoever. My contemporaries were ex-fighter pilots, bomber navigators and ex-naval officers. They flaunted their ranks and decorations on their bedroom doors: 'Flight Lieutenant Chinstrap, DFC and Bar'; 'Captain Pukka, MC.' I could hardly follow that by writing 'CSM Wilson, Whitgift OTC'! Diplomatically, I simply wrote 'Peter Wilson' on my door sticker and tried to grow up fast.

The first term was somewhat heavy going. The ex-servicemen had second-hand 'bangers' whilst I had only a rusty old bike. They played rugby, hockey and cricket very well; I only played tennis and, latterly, squash rather badly. They could recount War stories for nights on end; I could only talk about being an unofficial, unrecognised ARP runner. They could attract the women – both ex-service and ex-Women's' Land Army – whilst I had to do the best I could with the small number of young girls in the second year of horticulture. In those days, the 'agrics' and the 'hortics' tended to keep themselves somewhat to themselves and it was only after a term or two that territorial boundaries were broken down and the College began to develop a corporate identity.

With careful financial planning, and having learnt to keep my head down and 'go with the crowd', I began to enjoy College life more and more. The work was interesting, the staff were splendid and, as the year wore on, more and more social events and opportunities presented themselves. The ex-servicemen realised they were now at University and not fighting a war, with the result that I became accepted as a member of the College rather than being merely tolerated as a youngster in the company of an older, more experienced group. I started another YHA group and organised termly treasure hunts (on bicycles!), which proved quite popular. The hunt took place over a ten-mile course, with clues left in strategic positions every mile or so. The final destination was always a pub, with a few snacks laid on and free drinks for the winner. I made some close friends of students from my own age group and with one of them – a hortic student, Eric Cordell – I started a village youth club loosely attached to the village church.

The Wye College community was small and so everyone knew everybody. The staff extended a great deal of home hospitality to the students and we were on friendly terms with their wives as well as with our tutors. Working with the Wye Youth Club meant that I got to know the movers and shakers in the village community and by the end of the second year, life was great. I organised a 'Dunmow Flitch Trial', in which couples had to vie with one another and prove in court they 'had not had a wry word for a year and a day'. Several senior members of staff took part, William Deedes (Peterborough of the *Daily Telegraph*) and Dunstan Skilbeck acting as counsel for the couples and the 'Dunmow Flitch' respectively. In those days William Deedes was not so well known as he is now and I was lucky in securing his services by virtue of the fact that he was a close friend of Dunstan Skilbeck. (Skilbeck, I discovered, was a man with many important and useful connections). One pair of contestants in the trial was Professor and Mrs Mac Cooper. Mac entered into the spirit of the trial with gusto. The 'comic twists' were naturally preordained and well rehearsed. During the Mac Coopers' trial, the counsel for the Flitch queried whether they were legally married. 'Of course we are', said Mac, 'and here is our marriage certificate to prove it'. He duly produced a licence to keep a female dog for one year from the date stamp! The students fell about laughing and the village folk couldn't contain themselves. Seeing the College staff making complete fools of themselves was rather rare and really something to remember. To the delight of everyone, the Flitch was won by a popular village couple. The event was a sell out and helped to bring the College and the village a lot closer together.

Being a very close community meant that the non-academic members of

14. *Wye College Quincentenary Procession to Parish Church from Wye College, 1947. Author second in procession as Crucifer. Archbishop of Canterbury (Visitor to Wye College) in middle of procession.*

the College staff were as well known and respected as the teaching staff. One very key member of staff was the day porter, Aggie Cole, a spinster, who ruled over the territory of the Porter's Lodge with a veritable rod of iron. Everything was neat and tidy and woe betide anyone who tried to put a notice on the board in her Lodge without prior permission and approval. Aggie knew everything about the College and every member of it. She prided herself in recognising each new student within a week of term and this knowledge extended far beyond just knowing their name and degree course. There was no need for a 'rogues' gallery' of passport photographs, often seen in many large Colleges today. She was a confidante not only of the women but also of the male students and she kept up with them after they had graduated so that when any ex-students re-visited the College, their first port of call was to see Aggie Cole at the Porter's Lodge. Without hesitation, she would greet the returning alumnus by name and proceed to ask after their wife/girlfriend/children/career with unfailing accuracy.

Aggie was no beauty. Her clothes were never in fashion and she was small and dumpy, with grey hair tied back in an unruly bun. I doubt if she ever used make-up from one year to another but, beneath this somewhat unpromising exterior, she had a smile for everyone and a veritable heart of gold. We all loved Aggie Cole and if any new fresher unadvisedly said an unkind word against her, he or she would be sat upon with great alacrity. I doubt whether Aggie Cole ever had a day off work sick and she must have taken her holidays – if indeed she did take any – when no one was about, because she was never missing from the Porter's Lodge from one week to the next. How she found time to obey the calls of nature I will never know, as she had no relief back-up and was always to be found within two or three yards of her chair behind the telephone exchange.

When I revisit the College, as I do from time to time, there is always a different person residing in the Porter's Lodge. One might be asked kindly to sign the visitors' book but one would never expect to be known by name and I wonder if the Lodge occupants could identify half the staff, let alone all the students. Institutions like Wye College have lost something of great value with the passing of the traditional life-long servants, who, in their own unassuming ways, were as much a part of the corporate life of the College as the Principal and the Professors. But perhaps Julian Slade would bid us to 'Remind me to remind you, we said we'd never look back!'

There was a great deal of talent – and not just sporting talent – amongst my fellow students. One girl, Maeliosa Tohall, had a superb voice and could have become an internationally famous opera singer, had she not wanted to read horticulture. Another of the students, Michael Nightingale, was a well-read historian and archaeologist and he impressed us all by his excellent detective work in finding hidden treasures behind the ancient wooden panels of the 15th-century part of the College. To our eyes even more significantly, he discovered the whereabouts of the original 'Old Flying Horse' pub sign, which had lain unrecognised and forgotten in a village refuse dump.

Another outstanding character was John 'Beagle' Stephens. John acquired the nick-name 'Beagle' as he was responsible for starting a beagle pack at the College, building kennels for the hounds and training both the kennel staff and the student beaglers in the finer points of the sport. He was elected Master of the beagle pack and one of the lecturers, Noel Tinley or 'Tinners', as he was known, was the whipper-in. The pack had the full support of the Principal, who gave a lot of help and encouragement, including assistance 'in kind' in the form of a site for the kennels. He also underwrote the start-up costs of forming the beagle pack. On leaving College, 'Beagle' Stephens made a name for himself in agricultural journalism, and I wrote several articles for him when he was editor of *Span*, the house-journal for the agricultural division of ICI. (50, 84, 183).

Another notable contemporary was David Peto, an outstanding hockey player who formed his own successful seed merchanting company in Cheshire. We kept in touch with David and his wife, June, over the years and more especially when we lived relatively near to each other in Chester at the time when I held the position of Agricultural Director of R. Silcock and Son (see chapter 10).

Food and clothing were still rationed throughout my days at Wye, but we had a marvellous Domestic Bursar, Eleanor Knight, who did a number of deals with the College farm, so that a lot of food found its way into the kitchens 'off the ration'. Eleanor came to Wye from Swanley College and was a classic example of a tall, prim, but good-looking spinster who had complete command of any situation. If one wanted a dinner or a lunch for a special occasion, the 'special' epithet would be carefully vetted and, if it passed muster, a delicious spread would be laid on with all the trimmings. On being thanked, Eleanor would smile and say, 'That's quite all right, Mr Wilson, but I do hope there won't be too many events of the same kind this term or I will run out of rations!'

The ex-servicemen came to Wye with brand new demob. clothing, but those of us coming straight from school had to mend and darn with the help of mothers or girl friends. We had to wear gowns on formal occasions – official 'dinners in hall', examinations and church parades – and these regalia required not just money but precious clothing coupons. On one formal occasion when the Archbishop of Canterbury, in his role as 'College Visitor', graced us with his presence, a young newly-arrived demonstrator dared to come to the special College event wearing a Reading hood over a London graduate gown. None of the students knew the subtle difference, but Principal Dunstan Skilbeck was an expert on University regalia and was on to the poor chap like a ton of bricks. 'What the devil do you think you're wearing?' he demanded in full hearing of a large assembled company of academics, 'Pyjamas?' The chastened staff member, known thereafter by the students as 'Jamas' or – because he pressed on regardless – as 'Press On', mumbled an apology, but the story got round the College like wildfire.

Dunstan Skilbeck, being an ex-serviceman himself (he was a Wing Commander in the RAF), was particularly popular with the older College students recently demobbed from the armed forces. He was, however, strictly impartial and the College rules – designed more for those straight from school than for those from the battle field – were there to be obeyed. On one occasion he accepted the invitation of the Rugby Club to be guest of honour at their annual Club Dinner held in Canterbury. The College gates were closed at 10 p.m. each evening, but the Rugby Club event went on into the small hours. When the bus returned from Canterbury with the Principal in company with the noisy late night revellers, he duly bade goodnight to them at the locked College gate, rang the bell and said to the night porter, 'Be sure you take these fellows' names and have them appear before the College Secretary first thing in the morning'. They were all duly fined five shillings for late entry to the College, but the Principal had so arranged things that the fines found their way straight back into the coffers of the Rugby Club.

Dunstan Skilbeck recruited a very able staff, each member being committed to teaching his or her agricultural speciality to the very best of their abilities. The most distinguished academic was undoubtedly Louis Wain, at that time the only member of staff to be elected a Fellow of the Royal Society of London. Louis and his wife Joan were staunch supporters of student events and they both enjoyed attending activities such as the plays, concerts and dances organised by the students. Louis was an exceptionally able lecturer in his chosen field of plant biochemistry and he was in much demand for giving lectures to farmers and other audiences on the exciting advances being made in the field of plant growth-regulators. Louis in turn recruited Colin Mills, an equally able colleague, who taught agricultural chemistry related to animals. His special field was mineral metabolism and Colin is regarded as one of the world's leading authorities on the importance to farm animals of the elements copper and zinc. Colin eventually left Wye to join the staff of the Rowett Research Institute in Scotland. He became a Fellow of the Royal Society of Edinburgh and our paths crossed again when I returned to the UK and became involved with devising the nutritional standards for farm animals under the supervision of Sir Kenneth Blaxter, then Director of the Rowett.

Another outstanding member of the Wye College staff was James Wyllie, a canny Scot who had pioneered the analysis and compilation of detailed agricultural accounts for every major farming enterprise. James worked in this field long before the coming of the computer, and all the extensive clerical work involved in sifting through hundreds of individual farmers' accounts was done solely with the aid of an electric adding machine. James taught both accountancy and economics to both agricultural and horticultural students; his lectures were accompanied by a dour approach to his subject assisted with a dry touch of Scottish humour. Many of the lectures we attended 50 years ago have long since been forgotten, but the down-to-earth nature of James Wyllie's tutorials have lasted the test of time and are still recounted with affection at student reunions.

All the undergraduates were expected to engage in 'relevant and useful' agricultural work during the summer vacations. Indeed, the College tutors had a list of progressive farmers who were willing to take students on to their enterprises and at the end of the summer the Principal required us to produce a report outlining what we had gained from these periods of work attachment. In my first long vacation I worked with the system and was duly allocated a place on a Hampshire farm belonging to Captain Bomford, who was a progressive cereal grower and one of the first to install a grain-drier on his farm. I shared this vacation work with a contemporary, Phil Osborne, and we duly got ourselves down to Tufton Warren Farm on the Hampshire Downs. We drove combine harvesters for part of the time and, towards the end of our stay, worked in the grain-drying plant. By that time the grain had been harvested but needed to have its moisture content reduced to about 12 per cent before it could be safely stored prior to sale. The grain-drier was a bit of a 'Heath Robinson' affair and was very difficult to clean out, a job which needed to be done twice a week to prevent dirt and small weed-seeds choking the various ducts. It so happened that my tutor, Mr Tinley (or 'Tinners' as he was known), visited the farm to see how we getting on. Captain Bomford directed 'Tinners' to the grain-drier, which I was busily cleaning at the time, and called for me to come out as my Tutor wished to have a chat with me. I crept out of the grain-drier covered in dirt and dust and must have been an alarming sight, more like a miner than a farm worker! 'Tinners' was, I think, duly impressed that I was 'getting down to it' in no uncertain fashion. My other part-time duty was getting up early every other morning to move the poultry 'night arks' up and down the corn stubble. The 'arks' were movable poultry sheds in which the birds were confined every evening to protect them from predation by foxes. The arks were attached to movable poultry runs, which the birds used during the day. Both the arks and the runs had to be moved a distance of one length up and down the field. Movement down the field was no problem, but when one reached the bottom and had to pull the contraption up the hill on a parallel course, the work was very difficult. It was really a two-man job but it was made clear that the work was routinely done by a single farm labourer, so not to be shown up as a weakling, one naturally tried to do the work unaided. There were about a dozen arks and runs to move, and the whole operation took anything up to a couple of hours on the ascent, but only about thirty minutes on the downhill run.

On the other two holidays I decided to be a little more adventurous and find suitable agricultural work overseas. On one holiday I went to Northern Ireland, spending about three weeks on each of three different farms. The first was a very interesting establishment run by George Fulton, then President of the Ulster Farmers' Union. George made a name for himself reclaiming salt marshes from the sea. After two or three years of irrigation to wash out most of the salt, these were planted down to grass and clover pastures on which beef cattle were reared. George spent a lot of time with me discussing Irish agricultural politics, and he kindly took me to some farmers' meetings that he was attending. I found the placement with George and his delightful family very enjoyable and instructive. The next farm was a traditional dairy and sheep farm a few miles from Londonderry. The cows were Ayrshires and were milked by hand, the milk being used for a retail milk round on the outskirts of 'Derry'. The farm was in marked contrast to George Fulton's in every respect. The last farm of the trio was a company-run modern dairy, with all the newest labour-saving devices. The cows were milked in a herring-bone milking parlour, fitted with semi-automatic feeding machines. The milk yields and breeding histories were meticulously recorded and the overall management was very sophisticated, if somewhat top-heavy. All in all, the plans for this working holiday were most successful and encouraged me to do something similar, but further afield, the following year.

After a lot of digging and delving, I managed to find a contact in Sweden who arranged for me to spend the summer at a traditional Swedish farm at Hagby Gard, near Eskistuna. Getting there was a bit of a problem as money was very tight, so I decided to cross the Channel and hitch-hike my way across Europe, which in those days was legal and permissible and also less hazardous than it would be today. On the last part of the journey I was making my way up to central Sweden from Gothenburg and hitches were proving very difficult to get. Eventually a Danish car stopped and offered me a lift, which I gratefully accepted. After driving for about three hours we stopped for a tea-break and the couple had a long *sotto voce* conversation with each other which I could not follow. At the end of it the husband asked me where I was going to stay the night. 'Anywhere,' I replied. 'If I can't find a youth hostel, I will see if I can sleep at a police station!' It was then explained that the couple were on a one-week holiday which they had arranged to share with two friends but that these had, unfortunately, been taken ill the day before they were due to depart. As a result, hotel accommodation had already been booked and paid for, along the route I was proposing to travel. 'Would you like to take up our friends' booking?' they said, making it clear that they could not get their money back at this eleventh hour. Since it was all prepaid, I would not be expected to pay for anything except drinks. I had really fallen on my feet on this particular hitch, and travelled in unexpected luxury for the next few days. The farm, when I eventually reached it, was excellent, although very traditional and I learned a lot about Swedish farming methods and a little of the Swedish language. I also developed a close friendship with Karen, the daughter of the farming family who were my hosts for six weeks.

Because of the somewhat unusual circumstances in which I had been admitted to Wye, the Principal kept a kindly eye on me, albeit without a trace of favouritism. He would always stop if we passed one another in the quadrangle and pass the time of day. In my first year we lacked a Professor of Agriculture, the only occupied chair at that time being that of Horticulture. The agric. students grumbled a bit about this, as it seemed to imply a loss of status for them and Skilbeck was aware of this. One day he stopped me and said, 'Wilson, you can be the bearer of good news before I post it on my notice board. We will shortly have a Professor of Agriculture and he should prove popular as he has played rugby for the All Blacks and also has a blue from Oxford!' I had great pleasure in spreading the news around and within a couple of hours everyone in College knew that Professor Mac Cooper was on his way to Wye from New Zealand. The messengers of bad news have a reputation for being killed, but there was no doubt that this lucky break, which enabled me to be the bearer of good news, increased my position in the student pecking order quite appreciably.

When 'Mac', as he was universally known, eventually arrived, I put in a pre-emptive strike to get him to attend one of the monthly functions organised by the YHA group. I asked him to talk about the importance of foreign travel and its role in fostering good international relations. He readily agreed and it was a great privilege for me to preside over this talk, the first lecture Mac delivered in England after his arrival from the Antipodes. Mac was a gifted speaker. He always spoke without any notes but his lectures were all very well prepared. Although primarily research-based, his courses always had a practical edge to them. One always felt he was on the farmers' side and not delivering a scientific oration without relevance to the realities of farming. He was a tall, lanky individual, with an unruly mop of grey hair, which he used to push back from his face from time to time. Unlike most lecturers, he never spoke from the rostrum or behind the lecturer's bench, but instead preferred to sit in a somewhat ungainly manner on the students' side of the bench, which gave him closer contact with his class.

Coming from New Zealand, where land and sheep were cheap but the inputs to land very expensive, he preached the philosophy of good grassland management. He was, in a way, an early version of an organic farmer, in that he could not see the point of using large quantities of bought-in animal food or expensive chemical inputs in the form of pesticides and artificial fertilisers. He believed passionately in 'low-cost production' and claimed, correctly at the time, that 'British Agriculture was at half cock' because it was not husbanding its natural resources in the most cost-effective manner. In this respect he was a man before his time, for in the post-war years the prices farmers obtained for their products were high and so reasonably generous levels of inputs could be justified on their cost/benefit ratios. However, half a century later, when farm prices fell, the lessons that some future farmers had learnt from Mac Cooper in the post-war years stood them in good stead. As this book is being written, the economic conditions prevailing in the UK are much more similar to those then current in New Zealand, where subsidies were unknown and where farmers had to 'live off the land' and not off the tax payer. This is an important area, which is explored more fully in chapter 14.

From my point of view, Mac was indirectly responsible for charting the course of my career and for one very simple reason. Mac had studied and admired the work of Dr John Hammond, latterly Sir John Hammond. Hammond was a leading light in the world of animal science, the recognised authority on the growth and development of farm animals and also on many aspects of animal nutrition, animal breeding and genetics. Mac quoted Hammond extensively in his lectures, so that I came to admire the latter and his work through the excellent introduction and insight which Mac had given me. When I had passed my degree examinations, after two years study, I was required by London University to spend a further year 'in residence' before getting my degree. This posed a problem. An extra year doing what? The University had no advice to offer. I had managed to dodge the first year of the three-year degree, so it followed that I would have to add an extra year on at the end. What I did with it was my own business! Now there is not much that a 'semi-postgraduate' student can do in one year. The next degree would be a M.Sc., and this required two year's work 'in residence'. A Ph.D. took three whole years, which was all very confusing when all I wanted was to 'qualify' for my first degree. I gave the matter a lot of thought and decided to use the 'third year in residence' not only as the last year of my first degree but also the first year of an M.Sc.

With that decision duly made, I next had to decide what research work to do for the M.Sc. and this is where Mac Cooper's introduction to Hammond came in. Hammond had personally carried out, or supervised, a series of growth and development studies on sheep, pigs and beef cattle. There were a few loose ends which needed to be tidied up and as the Hammond growth programme had never been extended to poultry, I decided to plug both gaps and at the same time carry out more research in the Hammond series, but on growing chickens. Mac Cooper, not unnaturally, was quite enthusiastic and willing to provide the necessary resources whilst emphasising that I would have to work largely on my own, with minimal periodic supervision. It was also made clear that although my choice of poultry was, fortunately, less costly than a similar programme involving larger animals, Mac himself had no time for, or interest in, 'farmyard fowls'. He himself was essentially a cow man, with pigs as a second string. Poultry, so to speak, were 'for the birds'. Further, Mac had no laboratories in the Agriculture Department, so he had to negotiate with the Vice Principal, the formidable Norah Penstone, for space in the biology labs. This was my first experience of the cut and thrust of 'academic politics'!

At the beginning of my 'third year' at Wye I had to leave the close-knit society of 'living in Hall' and go out to the town of Wye and find myself some digs. This was relatively easy as I knew a junior member of staff who lived in the town and I also knew that his landlady was seeking another 'lodger' to complete her complement of three paying guests. I was duly interviewed by a large, busty, cheerful woman, Norah Houghton, and introduced to her small, attractive 19-year old daughter Diana. I duly passed the test. So for the next two years I was 'in residence at the University' but living in the town of Wye, with a downstairs bedroom of my own costing twenty-five shillings a week and the attractive advantage of the Wye Hill Cafe across the road, which provided all meals. These consisted of a full English breakfast, mid-morning coffee, scones and cakes, a three-course cooked lunch, mid-afternoon tea, sandwiches and cakes and an evening four-course cooked meal, all for the princely sum of thirty-five shillings a week. I returned there just recently and had morning coffee and scones for two. The bill for this light snack came to the equivalent sum of exactly thirty-five shillings!

Although technically a postgraduate on a London University postgraduate scholarship, my income was little changed from my first two years at College. The pennies had to be counted and there did not appear to be any easy way of augmenting my income. Students did not take on part-time jobs as waiters or barmen in those days, since Dunstan Skilbeck would quickly have put a stop to it; 'members of the College' (as he liked to call us) were not supposed to do casual work, thereby taking job opportunities away from the local population. But more and more demands were being made on my slender resources. I soon discovered that part of the deal with Norah Houghton was that her three lodgers were expected, in turn, to escort her daughter Diana to local Hunt Balls and other expensive activities of a similar nature. As the only postgraduate working in the poultry unit, I was expected to help keep the poultry staff 'happy', which meant buying them a round or two of drinks at the pub once every week or so. As I got more involved in village life and activities, additional expenses were incurred in 'keeping up with the village Joneses' and so I decided I simply had to earn a little on the side.

I spotted an advertisement in the *Kent Messenger* for a part-time lecturer in agriculture at the Technical School in Canterbury. The duties were not onerous – two half-days a week – and the stipend attached to the post was £100 p.a., about a third of my postgraduate stipend from London University. I applied for the job, was interviewed and got it, but imagine my discomfort when the following week I was summoned to the Principal's study and was asked to explain my actions. Skilbeck, it transpired, was a member of the Technical School's Governing Body and had watched my application going through the system. He did not object to my getting the job – indeed he might have put in a good word for me – but he certainly was upset that he had not been consulted. 'Did I not realise', he thundered, 'that I was still technically an undergraduate, completing my third year in residence, and as such bound by the rules of the College? Undergraduates did not lecture at other places!' I duly apologised profusely, asked him to put it down to my youthful inexperience and pledged myself never to apply for any other jobs without his prior permission. I was dismissed, duly chastened, but I was not forbidden to go ahead with the new job. In fact I noticed a broad smile cross Dunstan's face as I withdrew from his presence.

The Technical School job was interesting and taxed my ingenuity to the limit. The class I was given – agricultural studies – comprised the members of the School who had failed to get into more 'rewarding' courses, such as electrical engineering, mechanics, building construction and so on. The class was entirely male and they made it abundantly clear that they disliked being cooped up in classrooms to be lectured on stupid concepts such as Four-Course Norfolk Rotations, Carbon and Nitrogen Cycles and the elements of Soil Chemistry. They knew every trick in the book to make my job well nigh impossible. If I tried to use visual aids, they would put pieces of paper in between the power plug points so that the lights fused or never came on in the first place. They all 'asked to be excused' at least once every half hour and they were clearly failing to get any benefit at all from my efforts. After a couple of weeks I decided that, if I was to see the year out, something had to change and I spent a lot of time working out what form the change should take.

At the end of my heart-searching I went to see the Technical School Principal and asked him whether it was essential that I stuck to the syllabus I had been given, explaining that I found it very difficult to get the class excited about Carbon and Nitrogen Cycles. 'I don't care what you teach them, Wilson', he exclaimed, 'As long as you keep them quiet. It's the noisiest class in the whole School.' I thanked him and left, and then proceeded to put my plan into practice. I tore up the syllabus and concentrated on three subjects of my own devising: Sex, with reference to the reproduction of farm animals; Internal Combustion Engines, with particular emphasis on motor bike engines and Money, how to make it, keep it and expand it. The plan worked. I never had any more trouble with the 'difficult class'. Indeed I even had some boys asking to transfer into my course from some other far more worthy and useful subjects. When I offered to organise an outing to a stud farm, it was fully subscribed, even though it was in the boys' own time and partly at their own expense. At the end of the year I set and marked my own examinations, which all the students passed with flying colours. In fact, there was barely a detail in my lectures on reproductive physiology which they had not remembered! The result was that the Principal invited me back to help again the following year, and with a slightly increased salary (£110 instead of £100), although he never asked me for the details of my new and unofficial syllabus.

I have always been very grateful for this experience, which taught me the simple lesson that there are very few boys who are 'dead from the neck up', although there are many whose interest has never been kindled because of blind adherence to unimaginative and boring syllabuses. These young men would never set the world alight in the academic professions but, if well motivated, they could hold their own with their peers in jobs that were within their competence. Indeed, one of my perpetual worries when taking this class was that I would eventually find a boy who knew a lot more about the complexities of motor bike engines than I. My contingency plan was that, if such a thing ever happened, I would ask him to take the class while I acted as his assistant!

My research had to be fitted in with my work at the Technical School, my involvement in College and township affairs and with a fairly full social life. The work involved a lot of detailed dissections of growing poultry slaughtered at different stages of growth, having been grown on different planes of nutrition. I alone was responsible for the welfare of the birds, the making of their experimental diets, their twice-daily feeding and, at the appropriate time, their slaughter and dissection. The work was quite exacting and much of it had to be conducted in the evenings. I was allotted a corner of the Biological Laboratory for the bench-work involved, and, as already indicated, this laboratory was ruled over by Vice Principal, Dr Norah Penstone. Norah was, by training, an ecologist and a very good one. We had a reasonable rapport, in that my ecological knowledge was better than that of the average agricultural student, thanks to the excellent teaching of Cecil Prime at Whitgift, so by and large we got on well. However, Norah was a martinet; she ruled majestically over her biological domain, whereas I was to some extent an unwelcome interloper. Moreover she considered that I had been forced upon her by my supervisor, Mac Cooper, whom she did not take to as he was not 'a real scientist'. Norah worked late into the evening and would take an over-enthusiastic interest in seeing that I never overstepped any arbitrary boundaries she might have made for the running of the laboratory. On one occasion she stormed into the laboratory about 11 p.m. and asked when I was likely to finish. 'In about another hour or so,' I ventured. 'But you have got all the laboratory lights on', she said, 'and you are only working in one small area!'. I argued that every few minutes I had to leave my dissection and prepare some microscope slides in another part of the laboratory and so was constantly on the move. 'That's no excuse for wasting electricity, which comes out of my budget,' she exclaimed, 'Only have the lights on when absolutely essential!' Having won her trick, she marched out imperially, her long dark skirt switching behind her. Needless to say, once I heard her car drive away, the lights all went back on.

The research on the growth and development of chickens went well and there were no major hitches. I had to do all the work myself without any assistance, except for occasional help on the poultry unit in feeding and watering my experimental birds when I was away for some reason or another. I duly produced my four bound copies of my thesis, as required by the University regulations, and shortly thereafter was called up for my oral examination with my internal supervisor, Mac Cooper, and the external examiner, John Hammond from Cambridge. I was amazed at the attention to detail that Hammond had given to my efforts. Every page of my thesis was covered with hand-written marginal comments, all constructive and helpful for, as he said to me with a smile, 'You'll be wanting to publish this and it might help you meet the requirements of the editors'. In due course, I submitted three papers for publication (1, 2, 3) and the editor, not entirely to my surprise, was no other than John Hammond.

My four years at Wye passed quickly by and I enjoyed the experience of being an active 'member of the College' more and more. I have no doubt there were bad days and I am sure the weather was at times inclement, but one's memory selects and recalls only the good times and remembers them

15. Bunny, when we first met, Wye, 1949.

16. The author (third from left) and Bunny (third from right) in 'Aladdin' pantomime in the Wye Village Hall, Winter, 1950.

with great clarity. Early morning walks over the College farm before breakfast. Cycle rides with fellow students or with members of the Wye Youth Club into the beautiful countryside of East Kent. Long walks over the Downs looking for rare orchids. Lively debates in the common room about anything and everything. Quiet evenings in the homes of College staff listening to records or talking about the progress of science in their respective fields. 'Handshakes' with Principal Skilbeck at the end of every term to discuss 'How things were going'. Formal College ceremonies in Hall, Lecture Room or the local Church of St. Gregory and St. Martin at Wye. Graduation balls and formal dances, with everyone striving to look their best on slender finances and despite a dearth of valuable clothing coupons. Musical recitals, concerts and plays put on by teams of students and staff. Good, well prepared lectures, practical laboratory classes and weekly 'farm walks'. Playing games (as best one could) or supporting the College sporting teams. Festivities in the four village pubs. Walking hand in hand with a girl friend through the apple orchards when the fruit trees were in full bloom. Yes, there is no doubt about it, those were my Salad Days!

In those days I could read about a couple of general books a week, in addition to my text-books. Although I am a very slow reader, I have always enjoyed reading and my interests were very wide. At the beginning of my final, fourth year, my chosen author was Winston Churchill, whose *War Memoirs* I had acquired from money held over from a birthday present. The books – six of them – were an absorbing read, and I used to take them into the Wye Hill Cafe to peruse whilst eating my large cooked English breakfast. Reading required my undivided attention, but on one occasion I glanced momentarily upwards to find that a gorgeous dark haired girl with beautiful brown eyes had joined my table and was beginning to eat her cornflakes. Apologising for my rudeness, I introduced myself and learnt that my breakfast companion was a Miss M.E. Bunn, known to friends and family as Bunny, newly arrived in Wye as a junior Advisory Entomologist at the National Agricultural Advisory Service located on part of the College farm. Bunny rapidly replaced Winston Churchill as the centre of my attention. For the rest of the year, when not tending my chickens, teaching my Technical School students or running the Wye Youth Club, I was busily courting my future wife. We met in the autumn of 1949, got engaged on a boat in the middle of a lake in Regent's Park on March 13th 1950, and were married by Cyril Waynforth at Sanderstead on September 9th that same year.

And, as the story books say, we lived happily ever after!

CHAPTER 6

Edinburgh

'Science is nothing but trained and organised common sense,
differing from the latter only as a veteran may differ from a new recruit'

T.H. Huxley

IN THE 1940s, degrees in agriculture awarded by London University were all Pass degrees, Honours degrees being introduced much later. But many members of staff had Honours degrees in one or other of the pure sciences and not unnaturally argued that such attainments were a necessary prerequisite of an academic career.

It did not take me very long to realise that, with no money behind me and with a natural reluctance to take undue financial risks, a farming life was not for me. The next best thing was to try 'farming one step removed' by following an agricultural career in research, teaching or one of the agricultural support industries. Of the choices on offer at that time, a research and teaching career appeared to me the most attractive. The staff at Wye were happy, contented folk, enjoying a good quality of life and bringing up their families in idyllic surroundings. It was also apparent that they were not unduly stressed and appeared to have quite a lot of spare time to travel widely and develop personal and professional friendships in many different countries. This was clearly an attractive proposition. I wanted to join the Club.

How could I put my foot on the first rung of this ladder? My friend and tutor, Professor Wain, had the solution. 'What you need', he advised, 'is a good Honours degree. With that behind you, the world is your oyster'. 'How long will that take?' I asked, half knowing the answer. 'Four years' was the reply, although he pointed out that, with a little manoeuvring and a kindly Head of Department (and with a well-written letter from himself), I might be excused the first year, leaving only three years until the magic epithet 'Honours' was in my possession. But I was engaged and planning a marriage that same summer. How could I possibly contemplate working as an undergraduate for at least another three long years!

But Louis Wain's advice was sound. I needed that Honours degree, but I needed to get it in less than three years. So I did some research. Should the

60

17. The author on the occasion of his election as a Life Fellow of Wye College, 1993.

Honours degree be in Chemistry, Botany, Zoology or some more specialised field, such as Genetics? Writing round to the various Universities that offered these courses, I found only one that would give me an Honours Degree (or its equivalent) in one year – Edinburgh. Edinburgh had a clever system by which graduates from other Universities could take their final, Honours year in the Institute of Genetics, alongside Edinburgh undergraduates. They took the same final exam, which was graded, and the grades were directly equivalent to the Honours gradings of an Edinburgh undergraduate taking Genetics for his first degree. So back to the library I went and read up all I

18. The author and Bunny, Graduation Ball, London University, 1950.

could about Genetics, with special reference to Animal Genetics, as I had long since decided that I would specialise, if given the chance, on the animal rather than on the plant side of agriculture. After a month or so of detailed reading, I plucked up courage and applied to Edinburgh for entry into their Honours Genetics course.

I got called up for interview and, with my head crammed full of pertinent genetic information I duly appeared before the Buchanan Professor of Genetics, Professor C.H. Waddington, FRS. The interview went well; he did not seem to worry too much about my agricultural background or my wish to pursue an agricultural career, but he was most interested to find out what I knew about genetics. Luckily, one of the first books on genetics I had read was written by Waddington and as I had a good memory in those days, I knew much of his book almost by heart. He was most impressed. After a fairly long and pleasant but formal interview, he bowled a fast ball: 'What about your knowledge of statistics?' he asked, 'Because most of population genetics is very statistical'. I decided I had to come clean; honesty was the best policy and there was no way I could bluff my way through a barrage of high-powered statistical questions, so I confessed that this was my weak suit. I pointed out that statistics was not taught at Wye and that my mathematical education had stopped when I had taken Mathematics in my General School Certificate. I added, for what it was worth, that I had taken Farm Accountancy as part of my agricultural degree and implied that this was, at least in part, an exercise in applied mathematics.

Quite clearly, Waddington was unimpressed, but with a chuckle he admitted that statistics was not his strong suit either! He was working at the time in the field of embryological genetics and he argued that if he managed successfully to manipulate the early growth of a Drosophila fly so that a compound eye grew on its backside and not on its head, then that was statistically significant and the experiment had only to be done once to prove the point. So it transpired that Waddington did not regard statistics as a key part of the course and if one did well in the other sections, a lower performance in statistics should not stand in the way of getting a good Honours degree. I was informed there and then that I had been accepted into the Honours Diploma course in Animal Genetics and was then asked to make arrangements to move to Edinburgh so that I did not miss the introductory week, which started in a few days' time. With a light heart and an even lighter step, I found my way to the nearest post office and sent Bunny a telegram (which I still treasure) which read, simply:

'Purchase two kilts'.

Luck was on my side in more ways than one. Being married by then, our first need was accommodation in Edinburgh, and as my postgraduate stipend for the year in Edinburgh amounted to the princely sum of £260, it was clearly essential that Bunny found some suitable work to augment our income. (I have never understood the economics of the argument that two people can live as cheaply as one). I had taken time before my interview to meet as many of the senior staff at the Institute of Genetics as I could and one of these, Professor Alan Robertson, mentioned that he and his close colleague, Professor Jimmy Rendell, would shortly be looking for a technical

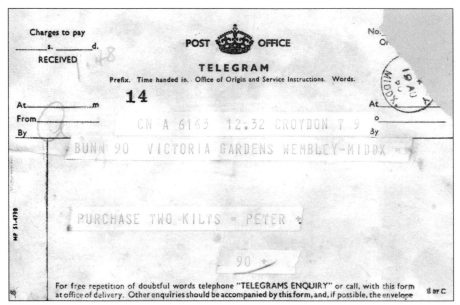

19. *The famous telegram, sent from Edinburgh, 'PURCHASE TWO KILTS', 1950.*

assistant to help with their long-term Drosophila breeding programme. He advised me to keep a close look at the 'Vacancies' section of *Nature*, as the post would shortly be advertised. In a couple of weeks this journal duly carried an advertisement for the position of technical assistant in population genetics, at a salary of £300 a year. This was a lower-grade post to her job as an Advisory Entomologist and would result in a drop in salary, but 'beggars can't be choosers'. Bunny applied, got the post, and our joint income overnight became £560, which was, in those days, a quite reasonable amount for a newly married couple. During that year Bunny was earning more than I was. She has reminded me of this fact many times during the course of our life together!

Again, seeking accommodation, luck smiled on us. My father knew an Edinburgh merchant who had a secretary who owned a flat in Morningside, which she only occupied at weekends since she lived and worked all the week in Peebles. After an exchange of very helpful letters with a very pleasant Scottish landlady, Miss Martin, it was agreed that we could have exclusive use of one bedroom and one living room of her flat and share the use of the kitchen. During the week the kitchen was ours and we were also permitted to use the second living room until Saturday afternoon, when the train brought Miss Martin back to Edinburgh from Peebles. She was an excellent cook and liked to spend most of the Sunday morning baking cakes

*20. The author (far right) with Bunny (second from right) and friends outside
the Institute of Animal Genetics, King's Buildings, Edinburgh, 1951.*

for afternoon tea, to which her friends were invited. They must have had
small appetites as there was always an abundance of good things left for us to
enjoy for the rest of the week. It was an ideal arrangement and Edinburgh
was a perfect place for two newly-weds to set up home for the first time. We
will always be grateful to Miss Martin for this very happy start to our
married life.

The group of students reading Genetics in my year were a closely knit,
interesting and friendly lot. About half of the team of 20 were finishing their
Edinburgh Honours degrees and the rest, like me, were attending the final
year in order to obtain an Honours diploma, having graduated in a variety of
subjects from other Universities; five came from overseas. We were, in many
ways, a small closed community within a large University, and in this respect
the 'family environment' was similar to that of Wye. Unless one was active in
sport (which I was not), the contact with students in other Faculties was
minimal. The Institute of Genetics had its own common room and small
refectory, and one could spend almost one's whole time within the confines
of the Genetics building unless one made a real effort to get out and about.
The number of people studying in, or working for, the Genetics Department
was about one hundred, only a little short of the 150 or so folk at Wye at that
time.

There were several postgraduate students in our group, and one was a
senior Agricultural Officer from Tanganyika, Harry Hutchison, who was

conducting research on the breeding programmes for indigenous African cattle. His posting in Africa was on the site of the much heralded, but unsuccessful, groundnut scheme introduced soon after the War by the Labour government. The scheme failed for a variety of reasons, but the chief one was that the assumptions about the mean annual rainfall in the Kongwa region were inaccurate. They were based on the records of a single rainfall gauge sited on the top of a small hill, the site of a mission school. These scanty records indicated that the rainfall was more than sufficient to grow groundnuts, but unfortunately on the plain below the mission station, where the groundnut crop was actually planted, the average rainfall was much less, so in most years the crop either failed or produced unacceptably low yields. However, although the rainfall was insufficient for arable crops, it was reasonable for grass production and eventually most of the groundnut scheme area was planted to pasture on which beef cattle were raised. Harry was interested in comparing different breeding strategies for the large numbers of animals required. He became a close friend and indirectly he must have influenced my thinking about African agricultural problems and the challenge of helping to solve them. It was an added advantage that, as a colonial officer on secondment to Edinburgh, he was entitled to a 'company' car, which was a magnificent new Citroen. We spent several happy weekends touring Scotland in Harry's car, and were also introduced to the delights of Indian curries, on which subject Harry was very knowledgeable. As a consequence we have become addicted to curry over the years.

Another post-graduate was an ex-serviceman by the name of David Wood-Gush. David had lost an arm during the War and at times he suffered a lot of pain from the short stump on his upper arm, although he never complained. I discovered that he was a keen squash player and agreed to play him at the game on a fairly regular basis. I thought that, minus one arm, his handicap would be such that he would have difficulty in winning against a two-handed player, but I was very wrong. David beat me almost every time and if I took a game off him, it was more by luck than good judgement. He had perfected the art of taking the centre-court position, forcing me to run around from side to side and end to end. Although I was faster and more nimble than David, I could never equal his skill at ball placement and invariably ended up by losing at least one set down. Later on, when Bunny and I were in Trinidad, we played bridge against another ex-serviceman who had also lost an arm during the War. He took a small cardboard box to bridge games and arranged his deck of cards in the little box. He was a very experienced and accomplished bridge player and, like David, rarely lost a match. We used to pull his leg by claiming that he had a fistful of spare aces in a secret compartment of his cardboard box.

Knowing that statistics was likely to be my undoing unless I worked hard at it, I went out of my way to pair up with a very bright Edinburgh mathematics student, who had transferred from the Maths Department to the Genetics Department for his final year. He had little knowledge of things biological and I had very limited knowledge of matters statistical, so we made a deal by which I helped him with his biochemistry and physiology and he assisted me with my statistics. We did a joint project together which was judged to be very good; he was congratulated on his grasp of biological theory and I was complimented on my knowledge of population genetics! This taught me an important lesson about co-operation, which has stood me in good stead throughout my career. The best work is generally done by teams, with each member contributing his particular skills and knowledge. Not much innovative science is now carried out by lone 'monks in cells', although let it never be forgotten that the principles of genetics were first postulated by Gregor Mendel, who was a monk, working away in his small cell at the Catholic Monastery at Brunn in Czechoslovakia.

Edinburgh was, in 1950, *the* place to study animal genetics. Under the inspired leadership firstly of Professor F.A.E. Crew, and latterly of Waddington, a team of internationally eminent research workers were assembled in the Institute of Animal Genetics on the King's Buildings campus. These included no less than nine folk who either had, or were soon to gain, Fellowship of the Royal Societies of both London and Edinburgh. The teaching was not narrow, as it is in many eminent Departments, where one particular aspect of the subject is pursued in depth. Rather it was very wide. The staff came from a variety of academic backgrounds: Medicine; Zoology; Physiology; Statistics; Embryology; Biochemistry – the list was endless. Waddington personified this multi-disciplinary approach through his own background and career. He started life as a geologist, which led him to consider palaeontology and the fossil record of our sedimentary rocks. This led him to consider the evolution of plants and animals, which took him naturally to the study of genetics. Not content with mastering traditional Mendelian theory and 'classical' genetics, Waddington interested himself in the development of animals from the point of fertilisation to the development of their adult state and so he mastered the subject of embryology followed by developmental genetics, working especially with the fruit fly, *Drosophila melanogaster*. Any detailed study of embryology calls into question the ways in which the different genes function at different stages in the embryological process, now causing the skeleton to develop, now making skin and hair to grow according to a definite and predetermined pattern. This in turn brings one into the realm of the protein biochemist, since the manner in which genes operate is by turning chemical reactions on and off, as required, at different stages of the growth process.

Waddington was the inspiration behind a very diverse, yet highly motivated, team and it was a privilege to be taught by these different masters of their respective subjects, bringing one right up to the frontier of knowledge in each specialised area. This distinguished group was eventually dispersed — many of the members making their mark in medical schools, biochemical departments, biometrics sections and many others. The Institute of Genetics is no more. The buildings are now occupied by the Edinburgh Engineers at Kings Buildings, but the work enthusiastically led by Waddington lives on in scientific Departments at home and abroad. His inspired teaching is remembered by dozens of his former students from a wide diversity of backgrounds, even including folk with an agricultural and nutritional training, such as myself.

Waddington not only inspired students in his own field of scientific expertise, but encouraged them to dig and delve in other subject areas. I felt a need to increase my understanding of physiology, so Waddington wrote a note to his colleague, Professor Witteridge, the Professor of Physiology in the Medical School, on the strength of which I attended first-year medical physiology classes for a term. Somehow or other I also became fascinated by philosophy. Another letter was written and I attended some of the first year philosophy lectures in the Arts Faculty. I found them fascinating but unhelpful. There was insufficient time allowed to discuss the subject with the lecturer involved, who unfortunately did not have the communication skills to make his subject come alive to an uninitiated science student. I wrote these lectures off as an 'interesting experience' but never got very much out of them. This was in marked contrast to Witteridge's physiology lectures, which were riveting stuff, keeping the students on their toes from the beginning to the end of the course. I got to know Professor Witteridge quite well and asked him why he gave so many lectures to the first-year students, instead of holding back his fire power to the final — Honours — year, when most Professors displayed their wares to a more senior group of students. 'Well, Wilson,' Witteridge explained, 'any fool can teach final-year students who already have a firm background in the subject, but it takes a more experienced person to lay the foundations of his discipline in the first year. That is why I always give pride of place to my Year One students. I only give the occasional lecture in the area of my own research to Honours students in their final year.' How wise he was, and throughout my academic career I have always tried to make it my business to put most of my effort at the front end of the students' course, leaving my colleagues to fill in the details once the basic foundation has been laid to the best of my ability.

Towards the end of his amazing career, Waddington turned his mind to an entirely different field and became a pioneer of the new subject of Human Ecology, interesting himself in social problems of a quasi-political nature. After Waddington embarked on these new endeavours, the work of the Institute of Genetics at King's Buildings changed. More and more of what used to be known as Botany and Zoology, each in its separate Department, has been developed into two new groupings known as Cell, Animal and Population Biology and Cell and Molecular Biology, all subjects embracing the different constituent parts of Biotechnology. In this sense Edinburgh is still an international centre of research excellence in the 'new' genetics. However, something of the intimate family atmosphere has inevitably been lost in the process, since there is no longer a single person leading the work in the same co-ordinated fashion that pertained under the paternal direction of Professor Waddington. Change is inevitable, and as genetics is all about biological change, I suppose it is natural to expect to see more evolution in the organisational structure in the subject of genetics than in any other area of science.

Waddington enticed many distinguished geneticists to Edinburgh in the post-war years and I was privileged to be taught by many of them. In addition to Waddington, Geoffrey Beale, Douglas Falconer, Charlotte Auerbach and Alan Robertson were all Fellows of the Royal Society of London; Eric Reeve and John Maule had been awarded Fellowships of the Royal Society of Edinburgh. Other internationally distinguished members of staff were Ruth Clayton and Mary Lyon, and I had been fortunate to have both of these outstanding geneticists as my tutors for part of my course. Most of these research workers were deeply involved in various aspects of fundamental genetics, but Alan Robertson, Jimmy Rendell and John Maule had applied their science to solving applied problems of animal breeding. They had all been instrumental in devising new and better methods of planning animal improvement programmes, not only in the UK but also overseas, and it was to such people that I was most strongly attracted. I never regretted my decision to come to Edinburgh for further postgraduate work.

Although Scottish Nationalism had not really attracted much attention or support during our year in Edinburgh, one event was by way of a taste of things to come. Halfway through the year some Nationalist-minded Scottish University students stole the Stone of Scone from underneath the Coronation Chair in Westminster Abbey, to deposit it a few weeks later in the grounds of Scone Castle in Perthshire. The event created a great stir, attracting support from a wide range of people, clearly indicating that Nationalistic sentiments were a latent force to be reckoned with. Although many pointed out that the theft was illegal and that the damage to the stone (which was broken during its journey to Scotland) was an act of vandalism, the general opinion was that the episode was very symbolic, setting down the principle that the Scottish nation was ancient and honourable and one which demanded pride and loyalty. The struggle between Robert the Bruce and the two Edwards was not over and done with and it was important for the politicians to realise that Scotland had been, and still was, a nation to be reckoned with. It was no surprise, therefore, when in 1998 the Scots voted for a degree of devolution with a revival of their own ancient Scottish parliament. It was also interesting to note that the vote in favour of this move was far greater in Scotland than in Wales, where the parallel referendum in favour of a Welsh Assembly was won only by a very slender majority.

Bunny and I joined various University Clubs during our year, the two most important being tennis and squash, but we thought we should take advantage of our time close to the Scottish mountains to learn how to ski. The Ski Club met regularly in the autumn and early winter, waiting for the snow to fall, during which time we practised using our skis by prancing around a gym simulating stops and turns on a very horizontal wooden floor! All this, we hoped, would pay off when eventually we set foot on an actual ski slope. The snows did eventually fall – in late January – so we took off for a skiing weekend at Killin, arriving at our bed-and-breakfast cottage late on Friday evening ready for an early start on Saturday morning.

There were no ski lifts in Scotland in 1950, and instead we had the use of a snow-mobile to take us to the top of the piste. The snow-mobile was a tracked vehicle which, when fully laden with well-clad skiers, was somewhat unstable on sloping ground. Half way up the slope the driver turned the vehicle too quickly and the snow-mobile overturned, luckily throwing us all out onto the snow instead of crushing us underneath the heavy tracks. What had been a joyful merry group turned into a sad-faced disappointed gaggle, who wended their way back to the cottage on foot, to dream about what might had been and to make a somewhat mournful return to Edinburgh. So we never learnt to ski and our later vacations in the Alps and the Scottish hills thereafter took the form of walking and scrambling holidays. We had the fun of seeing our children and grandchildren evolve into competent skiers, but we never progressed beyond the practice classes in the University gymnasium.

Our time in Edinburgh quickly passed. We both worked hard and even tended a small allotment near to the Institute of Genetics in order to provide a source of cheap, fresh vegetables. We learnt some of the finer points of Scottish country dancing. In our final month, when the examinations were over, we dressed up in our finery (literally top hat and tails in those days) and went to the Medical Students' Graduation Ball. At the end of our year, the British Association for the Advancement of Science held its annual conference in Edinburgh, with Prince Philip (as he then was) as its President. The Principal at that time was Sir Edward Appleton (the physicist who discovered the 'Appleton Layer') and his daughter was a well-known Windmill Girl at the London night-club of that name, which boasted throughout the War that 'We never close'. Sir Edward's daughter was summoned back to Edinburgh to help her father with the social side of the BAAS's programme, and we noticed, with amusement, that the most frequent dancing companion of Prince Philip throughout that week was Sir Edward Appleton's gorgeous daughter.

It follows from the above that our all-too-short year in Edinburgh had a major influence on me and on my subsequent career. Louis Wain was right to tell me that I should 'get a good Honours degree' and I like to think that I was right in spending this year reading Genetics under Waddington at Edinburgh.

When, later in my career, I was tempted back to Edinburgh, I rescued those two kilts, which had heralded the beginning of our first happy year together at 'Auld Reekie'.

CHAPTER 7

Uganda

'Give a man a fish and you feed him for one day;
Teach a man to fish and you feed him for life'

Chinese proverb

M Y YEAR IN THE Institute of Animal Genetics at Edinburgh was supported by the Agricultural Research Council (ARC). At the end of the year two options were open to me, the first was to serve in the armed forces for two years as a conscript, and the second was to pursue a career as a scientist in the ARC.

I was keen on the army option since I was disappointed that the end of the War had prevented me fighting for King and Country, as I had hoped and expected whilst at school. I therefore informed the local Army Recruitment Office that I was approaching the end of my academic studies and I was duly called up, firstly for a Medical (which I passed A1), and secondly to an Army Board who looked at my background and deemed that I was suitable material for a commission in the Army Education Corps. Now the Education Corps was not exactly 'top of the army pops' list; indeed most of the officers and NCOs were regarded by their colleagues in the fighting regiments as useless wimps! Somewhat unfairly perhaps, since an uneducated army would be a hindrance today.

Not wanting to be regarded as a wimp, I explored the second option and was duly sent by the London Office to the Poultry Research Centre (PRC) to be interviewed by its Director, a certain Dr Greenwood. The PRC was situated close to the Institute of Genetics in Edinburgh, so I duly walked across at an agreed hour and was given a lecture by Dr Greenwood on his strategy for his Institute. The main point of his oration was that the PRC had to do work 'ahead of its time' as the poultry industry was backward and unscientific and saw no need for scientific research. It followed that 'visitors' from the poultry industry were to be discouraged from coming to his Institute as they would not appreciate the work being carried out and might even argue that scientific research in the poultry sector was a waste of public money. He seemed completely uninterested in me, my background or the work I might be expected to do if and when I joined his staff and I left the

72

meeting firmly determined that this second option was not a real starter. All this took place in the Easter holidays between my second and final term at the Institute of Genetics. As I had plenty of work to do to ensure a reasonable qualification at the end of my studies, I put all thought of career development the following autumn to the back of my mind and contented myself with the day-to-day problems associated with my studies.

We had to count the pennies at Edinburgh just as I did at Wye and we could not afford a radio, or any other expensive household gadgets, but we did buy *The Scotsman* each day and also ordered *The Listener* each week, which gave detailed accounts of all the lectures which had been given over the air. One of these lectures interested me. It was about the 'Colonial University Colleges' which were springing up after the War in most parts of British Africa. The talk was by Sir William Hamilton-Fyfe, then Chairman of the Inter-University Council for Higher Education in the Colonies. Sir William's lecture extolled the virtues of the traditional academic subjects: Law (the Africans must appreciate good legal practices); Medicine (Africans must learn how to treat disease); History (the Africans must appreciate where civilisation had come from and where it was going); Classics (without which no man or nation could claim to be truly educated) and so on. I was furious! What an impertinence to assume that what was good for Britain was good for Africa! I put some note paper in my double key-board 25 year-old typewriter and sent Sir William a letter – I even kept a carbon copy! The letter, dated 8th April 1951, began thus:

'I have just read the report of your broadcast talk on, "Colonial University Colleges" published in *The Listener*. As I have lived in this country all my life, I know nothing at first hand about the conditions in the colonies. During my five years at University I have made many African friends from whom, and with the help of Conferences on African affairs, I have learned of some of the difficulties, and the opportunities, which confront our Colonies today.

'I have been particularly struck by the unbalanced ratios of students (in the various Faculties of our Universities) from the Colonies. This, I imagine, was the point of your remark concerning the numbers of doctors and lawyers studying here. As an agricultural graduate, I am particularly amazed at the paucity of agricultural and forestry colonial students at present studying in this country. At the last British-Africa Conference which I attended there were two agriculturists, fifty-odd lawyers and twenty-odd doctors, with a sprinkling of teachers and arts students. It seems to me that any country whose inherent wealth and basic industries are connected with the soil should seek technical skills in the agricultural and veterinary sciences before the academic skills in historical research or legal reform.'

I went on to say that I never saw advertisements for posts in the Agricultural, Veterinary or Forestry Faculties of the Colonial University Colleges and wondered whether any such posts existed. Having got that off my chest, I was prepared to forget the matter, and get on with my genetics revision, but to my surprise I received a telegram a few days later asking me to travel down to London by train to meet Sir William – all expenses paid. I had a most interesting and constructive interview with him, lasting several hours and including a good lunch, at the end of which I was offered the post of Assistant Lecturer in Agriculture at Makerere College, Uganda at the princely sum of £525 per annum. I sent a telegram to Bunny from London, which simply read:

Purchase two grass skirts

In July I obtained my Genetics Diploma with, surprisingly, First Class Honours. In August our first baby was born, and in October we boarded an Argonaut aircraft at Northolt airport bound for Entebbe, Uganda.

The Faculty of Agriculture at Makerere as it existed in October 1951 consisted of a portion of a store room with a collection of test-tubes, pipettes, burettes, microscope slides, and me. There were no students (and no microscope!). I was greeted by a member of the Government Department of Agriculture staff – Ron Stuckey – who was due to retire in a few days' time. He told me that our bungalow on Makerere Hill was not quite ready, so we were to stay with Bernard de Bunsen, the Principal of the College, in his official residence at the top of Makerere Hill. Ron offered to give us both what help he could in his few remaining days in Africa, and wished me well in my post, which, he thought, did not have much going for it. Up to that time all the agricultural teaching had been carried out by agricultural officers like himself on secondment from the Government. This arrangement was much better, in Ron's opinion, as these officers knew the country, whereas it was clear that inexperienced folk like me, working full-time on the University staff, did not. 'It won't matter very much anyway,' he said, 'as there aren't any students; but with any luck there might be a few in a year's time!'

Having driven us to the Principal's house, he turned to my wife and said 'It would be useful if you could turn up at my residence at 8 o'clock tomorrow as I have arranged for a line-up of potential servants for you'. 'How many do I need?' Bunny enquired, somewhat taken aback. 'Oh, I think you might get by with about six,' he replied 'although as you get settled in and need to do more things you may well find you need a few more'. 'How many servants do you have?', Bunny asked. 'Eight', he replied, 'but I really ought to get by with seven as we don't have any young children now'.

21. The author with Lassie our dog and daughter Rosemary on safari, Uganda, 1951.

So early the following morning Bunny arrived at the veranda of Ron Stuckey's house to be confronted with a long line of Africans bearing their *baruas* – hand-written references from past employers. These *baruas* were not very helpful. Most were very brief: 'Salimu has worked as my cook for six months. He leaves me as I am being posted to Nigeria', or 'Mzuwa has been my second houseboy for a year. He is leaving me as he wishes to obtain a more responsible position'. Bunny had quite a job to choose her staff from the two dozen or so lined up for inspection, but she obviously knew how to pick winners as they all turned out to be excellent and the head boy – Mugani – took it upon himself to teach the new *Memsaab* her job as head of an establishment on Makerere Hill. Mugani took Bunny under his sturdy wing and would run round the house after her saying 'Memsaab, you have left your keys in the door of the store. You must not do this because you cannot trust these boys!'

The servants were not highly paid but they were not inexpensive, especially to a person earning only £525 as an Assistant Lecturer. The Head

Boy and Cook each received about 80/- a month, the *Ayah* about 60/-, the Second Boy and Cook's Boy about 35/-, so the total muster roll cost us about £15 in cash each month. To this must be added the cost of their uniforms and, at the end of a two-year tour, their holiday money whilst we were on leave. It followed that the total bill came to something in the region of £200 a year – a fair slice of the £525, from which tax and pension provision had to be deducted. Many folk today would regard these wages as derisory, but often some of the servants, having received their cash at the month end, would solemnly hand most of it back saying, 'Please put it in the Post Office for me, Memsaab, so I can take it home when you go on leave'. Sadly, the PO clerks supplemented their wages by charging Africans for each deposit and withdrawal. Indeed, servants working for a regular wage were regarded as being fairly well off – a sort of middle class – by those toiling as labourers on African farms, who were paid almost entirely in kind – food and clothing and shelter – with little or no cash changing hands at all.

A few servants were badly treated, but not usually by British expatriates. On the other hand, some servants stole from their employers and often turned up late for duty after a night's heavy drinking. We were, however, fortunate throughout our six and a half years in Uganda. Our servants were loyal and trustworthy and when we left Makerere Hill after three or four months to go 'on safari', and thus had to part with our first cook (who had a family to support and a small local farm (*shamba*) to run), he insisted on inviting us to his mud-hut home to meet his wife (*bibi*) and children (*watoto*). He also presented us with some home-made craftwork as a farewell present. Bunny presented his family with some lengths of cotton cloth from which to make frocks and *kanzus* (long cotton shirts) and tears were shed by the women as they made their farewells. From our own personal experience, therefore, the relationships between the expatriate 'officers' (as we were all known) and their household staff were usually excellent, based on mutual respect.

Bernard de Bunsen – the Makerere Principal – was a tall kindly bachelor with an engaging smile, who came to Makerere from the post of Director of Education in Palestine. He was a very approachable person, much liked by staff and students alike, and he did a great job of transforming the former Uganda Technical Training College into, firstly a University College of London University and, later, into a fully-fledged independent University. These were years of rapid expansion. Student numbers quadrupled; halls of residence were built, new Faculties established, staff recruited, mainly from the UK, but increasingly from the ranks of the Africans who had passed successfully through College or who had qualified overseas. There was a great

22. The author with a group of Makerere students at Serere Experiment Station, Uganda, 1954.

sense of common purpose in the closely-knit academic community. The staff knew each other and most of the students well and there was much intermixing of African and expatriate staff which – in the early 1950s – was not happening in the government service, where almost all the 'officers' were from the UK. Most important of all, the University staff took their teaching duties very seriously, sometimes to such an extent that their own research suffered, with the result that, on their eventual return to the UK, they were virtually unknown and found great difficulty in finding suitable posts. This phenomenon was known as the 're-entry problem' and it resulted in many staff spending their later years teaching in UK secondary schools or accepting posts completely unsuited to their training and experience.

The Ugandans were happy and contented people. The country was well endowed with natural resources which were rich and productive. The climate was equable for most of the year as the geographical location near the equator was ameliorated by the altitude – most of the land mass being well over 3000 ft., with the mountains going up to 10,000 ft or more. The main cash crop was cotton and much effort was put into breeding new and better varieties which yielded well and which attracted high prices on the world market, dominated in those days by the Liverpool Cotton Exchange. The main food crops were green plantains (*matoke*), finger millet and sorghum (*wimbi*), with increasing amounts of local and imported maize, which was made into a sort of porridge (*posho*) by most of the town dwellers. Although locusts were a constant concern, there were effective control programmes in operation and an active locust research station, which gave early warning of locust invasions. There were large herds of cattle, sheep and goats, although the only animal exports – relatively valuable as a proportion of GDP – were skins and hides. Coffee and cocoa were subsidiary cash crops but 'cotton was king' and the government revenues depended heavily on the size and quality of the annual cotton crop.

In my first year I had no students and as my knowledge of tropical agriculture was zero, it was wisely decided that I should spend my first year 'on safari', seconded to the Government Agricultural Department, in order to learn what made Ugandan agriculture tick. Rather than embark on a number of 'Cook's tours', I argued that I should be more usefully employed by learning through doing, rather than learning through seeing. So in the course of my first year I was given a number of tasks, two of which taught me a lot not only about Africa but also about the interaction between the African farmers and the protectorate Government. One of these tasks was to answer a question that had bedevilled the Uganda Treasury, which was to improve the accuracy of the estimates for the size of the next Ugandan cotton crop. The traditional way of making these estimates was to delegate the job downwards to local (*Gombolola*) chiefs, who would in turn delegate it to some of their underlings. How these staff actually carried out the task was laid down by the Government, but not in a very precise fashion. They were taught to consider all the cotton plots as approximate squares or rectangles and to estimate the size of each plot by pacing both axes, thus getting an estimate of the area by simple multiplication – length x breadth. The assumption was that one pace equalled one yard. The chief himself, or one of his officials, then examined each plot and, from their local knowledge about cotton yields, estimated the weight of cotton which could be harvested from each plot. All the estimated plot yields would then be added up, thus giving an approximation of the tonnage of the crop in that area at time of harvest a few months later. The officials of the Treasury then added up the estimated yields from all the areas of Uganda.

Year after year the officials in the central government found that the crop estimates from the North of Uganda were usually over-pessimistic, under-valuing the cotton tonnage by 10 or 15 per cent. On the other hand the estimates from the East of Uganda erred the other way, and were invariably over-optimistic. As the amount of cotton planted in the regions varied from year to year (whether or not to grow cotton was the choice of the African farmer), it followed that the expected revenues of the central government were sometimes badly adrift. As everyone knew that reasonable care was taken in pacing the plots, and also that those responsible for the yield assessments from each field were pretty good at their job, why on earth were the estimates always so inaccurate?

I decided to watch the operation in the field and duly verified that it was conscientiously carried out. No corners were being cut here. However, it slowly dawned on me that the Africans pacing the fields were of very different statures. Those from the North were taller; those from the East were shorter. When I got out a tape measure and worked out the average pace length of people from the different tribes, the reason for the discrepancy was clear. The long striders were under-estimating the areas of their plots; the short pacers were over-estimating them. If each person involved in the laborious work of plot-pacing had his average pace-length measured, and if an appropriate correction factor was applied, then the estimated yields came out much closer to reality. This is a simple and elementary example of the need to observe what is actually happening 'down on the farm' and not to sit in remote offices devising ways of carrying out important tasks from a desk. For the rest of my life I have tried to spend time observing how things were actually being done 'at the sharp end'. Too many people have reached senior positions in many walks of life without much detailed knowledge of how things actually work in the field or factory or, for that matter, in individual University Departments.

23. The main campus, Makerere College, Kampala, 1954.

Indeed, some of the very experienced agricultural officers fell into precisely this trap. Crop rotations were practised in those days and the textbooks made out that there were three or four set rotations used in the cotton-growing areas. As no one ever challenged the text books, this folklore was handed down from officer to officer. I asked my students, most of whom came from farming backgrounds, if they thought that everyone managed their farms in such a simplistic manner and they assured me they did not. I then asked my African Assistant Agricultural Officer friends why they had not queried the text books from which they were taught. 'It would not be appropriate', they argued, 'to disagree with what the Senior Bwana had written in his book.'

Another task I was given in my 'practical year' was to make a detailed contour map of a newly acquired area of land destined to be transformed into an experimental farm, school and technical college administered by the Department of Agriculture. The terrain was in steep hill country, covered in dense bush. Although the selected site was relatively high, there were no vantage points from which to carry out some form of simple triangulation, nor were there any roads or cattle tracks traversing the area, although a rough dirt track passed round most of the perimeter. I had no idea about what to do, but clearly working out the contour lines needed a level, and measuring the farm obviously needed a chain. I duly obtained a Dumpy level (an instrument for fixing contour lines) and a staff, surveyor's chain and prismatic compass on loan from Government stores and was given the services of a local African Assistant Agricultural Officer. This officer had authority to recruit and pay a gang of 20 African workers armed with crude bill-hooks (*pangas*) to enable us to penetrate the bush.

The land in question was about a square mile in area, and the work was very slow. The hacking gang were able to chop down only about 100 yards of dense bush in twenty minutes, but it had to be done in such a way that they were making traces along the contour – not simply in straight lines. The weather was very hot – it was the dry season – and I found that we could only progress at the rate of about 900 yards in each 3-hour shift. After three hours the men needed a large drink of water, which meant two members of the gang being assigned to water-carrying most of the time. The nearest water-hole was over a mile away. After six hours there was a 'lunch break' followed by a well-earned siesta. For me the greatest thrill was circum-navigating the conical hill on the contour, and arriving back where we had started – with an error of less than two vertical feet. Although this relative success pleased me, none of the men, apart from my African assistant, could see the point of it at all. Tempers flared when we started on the next contour

– 250 feet higher up the hill. 'We've been round the hill once and ended up where we started. What's the point of going round again?' was the question put to me. My assistant explained to them that this was necessary in order to make a map. 'Why is a map necessary?' they asked, 'we've farmed this land for many years without a map; we've built our villages without a map; why on earth does the Bwana want a map?' Not speaking the language made it difficult for me to answer any of these pertinent questions – indeed I was not sure I knew the answers anyway! However, my wise assistant helped me out. 'If we split the gang into two and have a race as to which team completes its work first, and if we give a beer party to the winning gang, they may possibly see the sense in it!' he advised. 'But how can I lay contours with my Dumpy level in two places at once?' I enquired. 'You'll have to do the best you can running from one gang to the other,' he replied with a smile. Because of this, three or four weeks later, I was able to produce not only a tolerably accurate map of the area, but I was much fitter for my regular bush jogging, carrying compass, chain and Dumpy level as I went. Most important of all, my work gang completed their job without trouble of any kind and they were very appreciative of the beer party we enjoyed on the last evening, given in honour of the winning team. Bunny, however, was not very impressed. I arrived back at our rest house after dusk every evening, tired out and wanting only supper and a sleep!

There was a slightly unhappy sequel to this story. A passing Agricultural Officer was travelling in the direction of the government store from whence I had obtained my surveying equipment. He offered to return it on my behalf, thus saving me a long journey. I duly accepted his kind offer, handed over the equipment and promptly forgot all about it. Imagine my concern at receiving a formal letter some months later from the Government Auditor's Department saying they had evidence I had signed for a prismatic compass, Dumpy level, staff and surveyor's chain. All items of equipment were now accounted for apart from the prismatic compass. This was valued on the government books at £20, so would I please send a cheque for this amount by return? My error, I soon found out, was signing for the equipment but not obtaining a receipt from the officer who had offered to return it for me. By then he had left the country for good. I wrote out my cheque and had to cut back on my expenses for the next couple of months.

In 1952 Princess Elizabeth (as she then was) and Prince Philip paid a visit to Uganda and a special reception was laid on at Government House to enable the Princess to meet members of the Makerere staff. The visit was a grand affair, in the beautiful gardens of the Governor's official residence, and we all turned up in academic dress with the staff wives in their best bibs and

24. HM The Queen and HRH The Duke of Edinburgh at Makerere Garden Party,
Entebbe, 1954.

tuckers. During our first few years in Uganda Bunny made her own dresses,
as we could not afford the outfits available in the Kampala shops. Her outfit
on this occasion was a cotton frock with a much-used straw sun-hat, into
which she had woven some freshly picked Frangipani flowers. The outfit
looked very good indeed and I was very proud of her, but in the heat of the
afternoon the Frangipani flowers began to wilt somewhat and they looked
rather droopy specimens at the end of the proceedings. Princess Elizabeth
looked relaxed and charming; little did we realise at the time that, a few days
later, she would be told of the death of her father and informed that she was
now Queen Elizabeth II and no longer a Princess. She looked so young and
carefree in the gardens at Entebbe but the weight of high office has borne
down heavily on her shoulders ever since. We met the Royal family on
several occasions over the years; at Garden Parties at Entebbe and at
Buckingham Palace whilst on leave; when I received my CBE from her in
1986, also at Buckingham Palace, and dining with the Queen and the Duke
at Holyrood Palace in 1988. We were also invited to several Garden Parties at
Holyrood Palace in the following years. Her Majesty graciously presented
Royal Medals at the Royal Society of Edinburgh in 2000 (during the time
that I was General Secretary), so we have been privileged to have seen, or

been introduced to, her on many different occasions in various parts of the world. Having seen the Queen 'at work', we have a very great respect for the manner in which she discharges the duties of high office. In spite of meeting countless new people almost daily, she gives the impression that each person introduced to her is a very special individual, in whom she is deeply interested. It is sad that her family do not always come up to her own high standards, which set a very good example not only to the nation but also to the world.

The Governor of Uganda for most of our time there was Sir Andrew Cohen. He was a rare animal – a civil servant without any previous overseas postings appointed as a Governor by the Secretary of State for the Colonies. I have known several Governors in Africa and elsewhere and they have invariably been men of experience, wisdom and a willingness to learn from others. Cohen unfortunately lacked any of these qualities. He was inexperienced, clever but not wise and somewhat arrogant. He openly claimed that he was the only member of the Colonial Government 'to understand the African!' and this, not unnaturally, got him into difficulties. Halfway through his term of office as Governor he dismissed the Kabaka of Buganda, accusing him of halting the progress towards a more centralised system of government which was deemed to be a prerequisite of independence. This proved to be a great mistake and a few years later Cohen was forced to reverse the decision and welcome the Kabaka back from exile with all due pomp and ceremony.

Although perhaps the most left-wing of all the Governors of Uganda during the Protectorate period (for Uganda was never a true Colony), Cohen loved the ceremonies which went with high office, although he sometimes misjudged how these ceremonies should proceed. Soon after his arrival, on the occasion of a parade in Entebbe to mark the Queen's Official Birthday, Cohen appeared on the parade ground dressed in his full regalia. He was followed by the Kabaka dressed in the uniform of a Guards Officer, to which he was entitled, having been through Sandhurst and commissioned into a Guards' Regiment. The arrival of the Kabaka was the signal for the band to play *God save the Queen* but the Governor, with a broad smile across his face, marched across the parade ground in full view of the assembled crowd to welcome the Kabaka and shake him by the hand. The Kabaka knew better and, hearing the national anthem being played, came rigidly to attention and saluted the royal standard. This left a dejected looking Governor out of position, hand extended, not knowing whether to carry on with his greeting or stand to attention where he was. The senior African chiefs and officials present were not amused. 'What sort of a

Governor is this?' they asked, noting the way the parade was descending into a farce.

A story went the rounds of Uganda that when Cohen was on safari up country, he noticed that when a local chief returned home from an official visit, the police bugler attached to his residence blew his bugle to signal his safe return. 'I like this idea,' said Cohen, 'and in future, whenever I leave my Governor's rest house and when I come back, my official bugler will announce my departure and my return'. The Governor's wish was the bugler's command and this instruction was duly put into practice. Unfortunately, however, the first official visit of Cohen during his stay up country was to a senior Indian businessman's residence. He was, apparently, given an extremely rich curry for dinner and Cohen's constitution was not in tune with hot curries. That night Cohen had to leave his bed and visit the outside toilet (*kazi*) on several occasions. Although it was night, the bugler was at his post and on each trip to the toilet the bugle was duly played, with a repeat performance on his return. Cohen, speaking no Kiswahili at that time and having no interpreter, was unable to countermand his own orders and it was not until the following morning that his ADC was able to reinstate the *status quo*.

At the end of my first year on safari, my future boss, Professor Fergus Wilson, arrived at Makerere to take over the Faculty, with its limited collection of glassware. His immediate task was to build up the much-needed facilities for a brand-new and properly equipped Agricultural Department. Fergus was not a career academic – he had previously served in the Agricultural Colonial Service as an Agricultural Officer in Kenya and Zanzibar. Before coming to Makerere, he had been in residence at Cambridge on short-term secondment to the University to assist with the training of 50 or so colonial agricultural probationers. These were taking their Diploma in Agricultural Science before proceeding to Trinidad for their final year of study in tropical agriculture. Fergus was a slightly built, serious man with a good command of Kiswahili. He had very relevant East African agricultural experience (Makerere recruited students from all four East African territories in those days). He was an exceptionally hard worker and not afraid of doing things with his hands (thus he was soon given overall responsibility for landscaping the campus on Makerere Hill), but he was a little uncomfortable with his academic colleagues on the staff as he never knew when they were pulling his leg.

One of these colleagues, who arrived six months or so after Fergus had taken up his post, was Willem Boschoff, a South African who had been trained as an agricultural engineer. Willem's sense of humour was constantly bubbling over, much to Fergus Wilson's discomfort. On one occasion, prior to an important Faculty meeting, Willem gave prior notice that he would have to leave the meeting early as he had a very important engagement. When the day of the meeting arrived, Willem reminded Fergus of this and was informed that the point had not been forgotten. After about an hour Willem put his hand up, tendered his apologies and left the room, returning a few minutes later with a broad smile across his face and a bag of golf clubs slung over his shoulder. 'Goodbye, gentlemen,' he exclaimed, 'let me know tomorrow how the rest of the meeting went.' Fergus, like Queen Victoria, was not amused.

The ex-patriate staff at Makerere were small in numbers, they lived together in a closely-knit community and thus they all got to know one another extremely well. Although we only lived on 'The Hill' at Makerere for short periods in between postings on safari or up-country, we were, nevertheless, close friends of many of them, and we have kept up these friendships over the many years since leaving Uganda. Willem Boschoff was one of our closest friends and I have paid tribute to the lasting impression he made on my career in chapter 18. Another good colleague was Kenneth Ingham, the Professor of History, and we were entertained many times on the Makerere campus by Ken and his charming wife, Elizabeth. On returning to the UK, Kenneth became the Director of Academic Studies at Sandhurst, moving on from there to take up the chair of African History at Bristol University. Kenneth is a world authority on East African History and he has written many books on the subject, which are widely referred to by workers in this interesting field. Another outstanding member of the post-war staff at Makerere was Margaret Macpherson, Lecturer in English Literature. Margaret produced many plays during her time, mostly as part of her students' programme of work. Some, however, were organised by her for the members of staff for their own pleasure. I remember in particular a very professional performance of Hamlet, with the University Registrar, Paul Vowles, playing the leading role. The close relationship between colleagues on the Makerere staff was reminiscent to me of the academic community at Wye. It made me realise that there must be a time in the development of a new institution when the staff is sufficiently small to enable this close bonding to take shape, with benefit to all concerned, not least the students. Large Universities, like London or Edinburgh, do not (and probably cannot) have anything like a comparable community spirit and they are the poorer for it. At both Wye College and Makerere this common purpose grew and prospered, and it was very exhilarating being part of the action at that time.

The acquisition of the Makerere University Farm at Kabanyolo and the building of halls of residence and classrooms to enable students to spend their practical year nearer to the University campus, took time, but at the end of my first year Ron Stuckey's prediction was proved correct and six students enrolled to take the new full-time agricultural course. Fergus (still at Cambridge) quite rightly deemed the practical preliminary year to be important, as the majority of the students, although coming from a rural background, had never milked cows in a milking shed or had any experience with anything other than the most primitive tools and machinery. He therefore arranged for me to be stationed up-country at a Government Experiment Station at Serere, in the Eastern Province, in order to supervise and teach the first-year students during their preliminary year. The station was well equipped and part of the deal was that a staff house was made available for my small family and suitable accommodation was provided for the students. I was the tutor-in-charge, and it was my job to marshal a teaching team from the staff of the station and to arrange for practical work to be carried out by the students on the various sections of the station in rotation. In those days the European quarters were sited on the high ground of the station, near to the offices and laboratories, whilst African staff (and my six students) were accommodated one half mile away on lower ground in the African quarters. I was not too happy about this arrangement, but I had to accept it as we were guests on a Government-run station, and official rules and regulations had to be obeyed.

The students needed to be fed and their 'hall of residence' needed to be kept clean so I employed local Africans to act as cook, cook's boy and houseboy. The students, however, were not happy. They came not only from different tribes but also from different parts of East Africa. In that first year I had three students from Uganda (all from different parts of the country), two from Kenya and one from Tanganyika (as it was then called). They all had different dietary habits and the local cook failed to provide an acceptable service. I realised that unless I did something about their food they would be a discontented lot and this would bring them into conflict with the senior staff of the station. I therefore asked the students to elect a 'Hall President', who would, within reason, be responsible for giving the cook instructions and also for ordering rations from the local village store (*duka*). The first student elected to this office failed to deliver a satisfactory service, so a new election was held and another student took over these delicate responsibilities. I decided it was a lot of additional work for any one student and arranged that each student would take the job for only six weeks and then hand over to someone else. All students therefore had to take a turn as Hall President. This worked well and I heard no more complaints about the food. Indeed, such was my confidence with the way things were working out, that I instituted monthly 'high tables' when I dined with the students and brought along an important guest. The local District Commissioner graced the high table on one occasion and the Chief of Police on another. The students took great pride in laying on a special feast on such occasions, for which they were rewarded with an appropriate increase in their messing allowance. With only six students and three staff, my budget was stretched to the limit and I repeatedly requested Fergus for additional cash. He replied that there was a set allowance for all Makerere undergraduates which I was duly and properly receiving and it was up to me to make ends meet. I argued that an allowance per person for six students could not be equated with an allowance per person for a large hall of 200 students. Eventually I won my case and my domestic budget was increased by 10 per cent.

Many years later one of those six students came to the UK to take a Ph.D. degree at Bangor University and I went to Wales whilst on leave so see how he was getting on. He was doing very well and obtained a good doctorate. When I asked him to reminisce about his year at Serere he said, 'One of the best things you taught us was how to take responsibility for our own well-being. We may well have forgotten a lot of the lecture material but acting as Hall President was a lesson none of us ever forgot and it has stood us in good stead throughout our careers'. I was very touched by this, but on reflection it indicates that most folk will rise to a challenge. They would prefer to be in charge of their destiny, even if they make mistakes, rather than to be treated as subordinates under the direction of someone divorced from the group. This applied just as much to political independence for former colonial territories as it did to taking responsibility for pay and rations in a small student hostel.

Money was tight throughout our time in Uganda and we could not afford expensive hotel holidays. The days of the 'package holiday' had not yet arrived, although the game parks of Eastern Africa were slowly being opened up to visitors from overseas. Our first 'holiday' was a safari to the Murchison Falls in Northern Uganda, which required a long drive on rough roads to a small landing stage on the Nile. Here we boarded a ramshackle boat, similar to that shown in the film *Africa Queen*, and proceeded slowly up river to another landing stage a mile or so from the Falls. At this point we disembarked and trekked the rest of the way through open country, teeming with lion and elephant, accompanied by armed rangers, until we finally arrived at the top of the raging torrent and stood on a rickety wooden bridge which spanned the Falls. This holiday must have lasted only about a week but it was very memorable, firstly because it was the first time we had

*25. Rosemary and David outside our College House
at Serere Experiment Station, Teso, Uganda, 1954.*

been on foot safari in a game park and seen so many wild animals at close quarters, but secondly because the view from the crest of the Falls must be one of the best sights in the whole of Uganda.

Subsequently we spent several very happy annual holidays in Kenya, on the top of the Great Rift Valley. These holidays were spent on a 'White Settler' farm near Kabete, and the story about how we came to know the family who ran the farm is an interesting one. Our Serere bungalow was adjacent to the residence occupied by Jock and Osla Low and on one occasion the Lows had gone north to spend a week with their son, who was an Agricultural Officer working in the West Nile District. Whilst away from Serere, they lent their house to some Kenyan friends, Ron and Nan Pleadon, telling them that if they needed anything at any time they could call round to the Wilsons who lived nearby. In the middle of the night after their arrival, one of the Pleadon's daughters woke up with severe stomach pains and Nan Pleadon rightly concluded that she was suffering from acute appendicitis.

They followed instructions and came to wake us up, asking what should be done. Now the nearest hospital was at Mbale about seventy miles away, so I hastily put on some mosquito boots and drove next door to collect the young girl and her parents and drive them to Mbale. The journey took over two hours and must have been agony for the poor girl, who was lying on the floor of my Standard Vanguard van. On arrival at Mbale at dawn, the diagnosis of acute appendicitis was confirmed, but it was deemed best to call for an air ambulance and fly her to Entebbe for the operation. This was successfully performed and the young girl recovered. Her parents were most grateful and we were persuaded to visit them on their farm in Kenya for several holidays which we much enjoyed. The Pleadons eventually sold up in Kenya at the time of Independence and moved to Australia. We had kept in regular correspondence with them both before and after the move. Some twenty-five years later we revisited them in their retirement home near Melbourne. The daughter, by then married with a family of her own, vividly remembered her rough drive through the night and graciously thanked me for saving her life. This was clearly an exaggeration, as the bumpy ride in my van could easily have ruptured her appendix and killed her. What she did not remember was that I arrived in Mbale dressed only in pyjamas and mosquito boots, much to the amusement of the townsfolk, who must have thought I had taken leave of my senses!

For most of our four years at Serere the Officer in Charge was Merryn Watson. Merryn was, in many ways, similar in appearance and personality to Dunstan Skilbeck, the Principal of Wye College. He was of medium height and sturdily built with a pronounced public school accent. He was authoritative to the extent that he was held in awe by many of his junior staff, and he made sure that everyone knew who was in command and jumped when ordered to do so. He was a bachelor and the only folk who were able to handle him were his household staff, who knew how to exploit some of his eccentricities.

One memorable evening Merryn Watson organised a dinner party for his station colleagues and their wives, with the District Commissioner and his wife as chief guests. He had instructed the cook to roast two large chickens for the occasion, thinking that they would be adequate. Unfortunately, his instructions were slightly ambiguous. Although it was clearly understood that two birds were to be cooked, it was not made clear that both chickens were to be served up! The first chicken arrived, and Merryn deftly carved it into a sufficient number of portions to provide meat for all the ladies present. Pleased with his efforts with the carving knife, and seeing that every lady was adequately supplied, he then ordered the house boy to bring in the other chicken. 'There is no other chicken', was the reply 'You told the cook to roast the chicken, which he did'. The other bird had been eaten by the staff! As time did not permit of the preparation and cooking of a third bird, Merryn had to go round the table and remove one half of the portion presented to each lady, in order to feed their hungry spouses. Merryn did this with great aplomb, as though it was a common occurrence. When the dessert was produced no plates were provided, so he promptly served the first portion directly on to the table saying 'That will teach them to eat my second chicken!'

Merryn's authority permeated all aspects of life at Serere. Although the expatriate staff numbered only six officers, three of whom were married, the station had a swimming pool, two tennis courts, a nine-hole golf course and (built by the European staff by their own unaided efforts) one squash court. The 'rule' of the station was that all the staff, and their wives, played tennis three times a week. On Sundays the station was 'Open house' to government officers from miles around, the attraction being 'tea and drinks by the swimming pool', with the expectation that the three wives would preside over the preparation of piles of sandwiches and home-made cakes. If any staff failed to turn up on any one of these four important weekly occasions, they were presumed to be sick, and therefore Merryn took it upon himself to make an official visit to their home to enquire after their health.

On one occasion it was a wife who was 'absent without leave' and Merryn was concerned to find her confined to bed. He insisted that, first thing the following morning, she visit the local dispensary, presided over by an African Assistant Dispenser (Grade 3). This instruction was duly carried out, and the Dispenser, unaccustomed to dealing with European wives, was agonising over the problem of how he could relay the results of his brief medical examination in the correct manner. He decided that the written word was preferable to the spoken one, particular as the wife in question had only a limited command of Kiswahili, so he laboriously wrote out, in beautiful copper plate, a letter on official government medical department notepaper addressed to her husband, instructing her to take it directly to him. The wife rushed home with the unopened letter, fearing the worst. Her husband tore it open and read (when translated) 'Dear Sir, I have the honour to inform you that your wife is suffering from a slight touch of pregnancy. I am, Sir, your obedient and humble servant, John Ikara, African Assistant Dispenser (Grade 3)'. There was much merriment and congratulations at the bar by the swimming pool that evening after the winning game of tennis had been played.

Merryn was an expert biologist, with a particular expertise in Zoology.

His two main forms of relaxation after a hard day's work were reading detective stories (he got through about three paper-backs a week and then donated them all to the 'station library', which was a large bookcase behind the bar in the club house) and collecting and studying small mammals. He became the Uganda authority on small mammals and published many papers on the subject in the *East African Agricultural Journal*. He was also knowledgeable about the botany of Uganda and particularly of the area round about Serere. My early school-days' coaching in taxonomic botany from Cecil Prime encouraged me to team up with Merryn on plant collecting expeditions, and as a result of this joint interest we published a paper on the flora of a nearby village (5), which I was studying in depth from an agricultural angle.

As mentioned earlier, the standard textbooks on Ugandan Agriculture described a small number of different crop rotations, implying that these rotations were invariably followed by all the farmers in a prescribed area. My own observations and my students' comments indicated that this was not so. In fact, very many different rotations were practised, the actual choice being made for a number of varying, but good, reasons. I thought it would be a useful exercise for the students to spend about ten days living in one of the nearby villages, talking to the farmers and observing all their farming operations and visiting all their fields. By close inspection of the ground, it was possible to detect with reasonable accuracy not only the plants being grown at the time, but at least the previous two crops, since the progeny of these latest plantings, or 'pioneers', could be found in the field for several subsequent years. The students enjoyed the experience and they learnt that there was a lot of useful knowledge to be gained by talking to farmers at length about their cropping decisions and the reasons behind them. As potential agricultural advisers, it was important that the students should realise that, not only had they some valuable expertise to impart, but they in turn had much to learn from the farmers they were helping. These village surveys took place during the last two of my four years at Serere, and were duly written up in conjunction with Merryn Watson. Merryn acted as our 'consultant' as he spoke the native language, which, unfortunately, I did not (4, 10).

Underneath his gruff exterior Merryn was a kind and generous person and we became close friends. He was godfather to our eldest son – David – and took his responsibilities very seriously. A year never passed by without David receiving a card and a present on his birthday and also every Christmas.

With only three staff wives on the station, social life was somewhat restricted. We all entertained each other to dinner parties in rotation, but in such a limited community conversational topics were few and other ways of passing the time had to be discovered. Our neighbours, Jock Low and his Norwegian wife Osla, were keen bridge players. 'Keen' did not imply 'good' but they knew the rudiments of the game and were happy to teach us. We were willing pupils and we developed a liking for friendly bridge which has stood the test of time. We discovered that the members of a nearby Catholic Mission, run by White Fathers from Holland, also enjoyed playing bridge and so we occasionally entertained two White Fathers to dinner and bridge at our Serere residence. The White Fathers were delightful bridge companions, kind and courteous and very knowledgeable, with that rare knack of being able to tell you where you had made mistakes without appearing unduly critical. At the end of our first bridge evening together, the Fathers said: 'You have won by 900 points. At 1ct per hundred here is our 18 cts-9 cts each!' (There were 100 cents to the East African shilling, which in those days was more or less equal to one British shilling, or 5p in today's currency.)

Soon after we had held one of these bridge parties at Serere with our two White Father friends, we learned that one had died the following week of acute cerebral malaria. All expatriates took anti-malarial drugs regularly but although these successfully warded off most attacks, some of the malarial parasites still managed to get through. We suffered from malaria several times, but thankfully neither of us were smitten by the dreaded cerebral variety, which sent the sufferer to an untimely grave within hours rather than days.

When playing bridge with strangers, one could not take it for granted that 1ct per hundred was the universal going rate. On another occasion I was asked to play tennis for the Serere station team against the local Indian tennis club in Soroti – 20 miles away. We were soundly beaten at tennis (the Indians had a club of over 100 members from which to select, whereas we had only six male players at Serere). After the match we were asked if we would like to go to the bar for drinks and a rubber or two of bridge. Anxious to find out whether my newly acquired skills were up to the mark, I accepted, but my companions sensibly declined and went back home. Some three hours later we finished the rubber, and I found to my surprise that, through a lucky run of the cards, I was 3000 points up on the evening. My Indian friends were a little taken aback but, being good hosts, they said 'We have, of course, been playing for twenty shillings a hundred points, so here is our cheque for six hundred shillings'. To me, six hundred shillings was a small fortune so I demurred, but was told that it would be grossly impolite to decline, and, of course, had I lost I would have been expected to pay up. I returned home a richer, but also a wiser, man and ever since I have made it my business never to play bridge with strangers unless I know the stakes and agree them in advance.

Merryn Watson was eventually promoted to the post of Director of Agriculture in Somaliland and we have kept in touch with him, not only when he was working in the Horn of Africa but during his subsequent retirement to a country house in Co. Durham. His job as Head of Serere Experiment Station was taken by Dennis Parsons who, like Merryn, had served as an Army Officer during the War, and who also, like Merryn, had been awarded a MC for his valour in the field. Dennis was a family man and his wife Kath was a welcome addition to the small female establishment at Serere. Unlike Merryn, Dennis was a keen sportsman and arranged for the male members of the Serere community to go on numerous shooting expeditions. Whereas Merryn preferred studying the wild life of Africa, Dennis liked to get them safely shot and tucked into his game-bag! More contemporary with me in age were the specialist 'Research Officers' posted to Serere. A leading character was Walter Hirst, a down-to-earth Yorkshire Botanist who did not suffer fools gladly. On one occasion Walter had no hesitation in telling the Governor (when he paid us a visit whilst on safari nearby) exactly what was wrong with the way the country was being run at that time. Another extrovert member of staff was Dick Horrell (always known as 'Cass', an abbreviation for 'Cassanova', a nickname which he had acquired at Cambridge), who was a pasture agronomist. These younger men on the station were bachelors and as there were no unmarried women officers within a hundred or so miles, their social lives were somewhat restricted!

John Hammond – my Cambridge-based mentor – on hearing that I was going to Africa as a University lecturer, gave me some sound advice before I left home. 'Africa,' he said 'needs a lot of practical help, most of it elementary even to the point of being obvious. For instance, many African farmers do not appreciate that farmyard manure could double crop yields. It is important that you point this out to them. However, if you content yourself with such simple mundane matters, you will be quickly forgotten and you will find many doors closed to you when you eventually come back to this country. What you must do, therefore, is carry out some original research, which you must publish in reputable journals, so that on your return you will be known by your published papers, which will last the test of time.'

I was fortunate to have been given this advice and I have always been glad that I not only remembered it, but put it into practice. Soon after taking up my duties at Serere, I worked out a project which, like my poultry work at Wye, sought to extend the studies on the effect of planes of nutrition on animal growth. The work carried out by Hammond, and latterly by me, had worked on whole animal carcasses, but animal bodies contain varying amounts of fat, and this fatty tissue is capable of being laid down and reabsorbed according to whether the animal is being well fed or semi-starved. In order to ascertain whether vital tissues and organs have been affected by different planes of nutrition throughout life, it is therefore necessary to correct for these differences in fat content, in order to compare like with like. So I embarked on a series of experiments using as my material the East African Dwarf Goat.

This work, which led to my Ph.D. degree, had to be done in my spare time. Merryn Watson was about the only member of the Serere staff who had any appreciation of what I was doing and he helped me by providing goats and laboratory equipment. Fergus Wilson, as my professor, would normally have been expected to supervise my work but, knowing nothing about the subject (as he was primarily a crop husbandryman), he kindly arranged for me to be supervised by John Holmes, the Professor of Physiology in the Makerere Medical School. John seemed to think I could get by without any supervision on his part, but for the three years I was working on my project, he came up to Serere twice a year to see how I was getting on. He always brought a shot gun with him and after he had spent a convivial evening in our house, he would say 'Right, Peter, where shall we go shooting tomorrow?'. It was soon apparent to me that provided I took John to some good duck shooting locations, I would not have to explain the intricacies of my agricultural work in medical terms. After John's visits our larder (no refrigerators in those days!) was packed full of ducks, geese and guinea fowl, most of which were too tough to eat anyway, and all of which had to be cooked within a few hours of my return if they were not to putrefy beyond redemption.

If ever there was a Ph.D. done with the crudest of laboratory equipment and the minimum of help and advice, this was it. I wrote the work up whilst Bunny was home on special leave to have our second baby. A few months later I once more submitted the statutory four bound volumes of my thesis to Senate House, London. As previously, the oral examination was conducted by Mac Cooper and the great John Hammond and once again it was a very pleasant affair. Hammond handed his annotated copy of my thesis back to me, with 'some suggestions for improvement prior to publication'. This time Hammond thought that the work was of sufficient importance to warrant a short paper in *Nature* (7) in addition to a series of papers in the *Journal of Agricultural Science (Cambridge)*, which he edited (9, 11, 15). When the work was re-written for publication, Hammond helped me with the meticulous checking of the references, as by the time the work had reached that stage, I was back in Uganda and remote from a good reference library. John Hammond was indeed a very great man and many folk, like me, will be for ever in his debt. In contrast, I do not recall Fergus Wilson taking any interest in my research, or even asking to see the eventual outcome.

I wanted to publish some of my research on goats in a way which might be of use to general agriculturists working in East Africa, as I realised that the growth and development work was only of interest to basic scientists. I therefore published some work I had carried out on goat behaviour and feeding preferences in the local *East African Agricultural Journal* (8), since it was of importance to know what plant species browsing goats preferred. I also made a study of the records of the Serere dairy cattle herd and published them in the same way (6). It has been amusing to find that the last two publications are still referred to by agriculturists working in East Africa. When being introduced to new workers in this field, it is interesting when I am greeted with such words as, 'Oh! You must be Wilson the Goat!'

Our first few journeys to and from Uganda were by plane, but we also travelled sometimes by sea, on the Union Castle Line. After our first home leave we had arranged to travel back to Uganda on a Comet, which was just beginning what was hoped to be a long era of jet travel. As luck would have it, a week before we were due to fly, I received a letter from Fergus Wilson asking me to do a small job for him in Cambridge, which would take a couple of days. This meant that I had to cancel the Comet flight and transfer to a slower aircraft, which disappointed me greatly at the time, as we were both looking forward to experiencing this exciting mode of transport. It so happened that the Comet we had been booked on crashed into the sea off the coast of Italy. The plane was, apparently suffering from 'metal fatigue' in spite of its low number of flying hours, and the dawn of the 'Jet Age' was delayed for several years as a result of this tragedy. I have never been a fatalist nor do I believe in pre-destination, but we both felt that someone was watching over us on this occasion. It reminded me of a contribution to the war-time 'Brains Trust' radio series made by C.E.M. Joad. In this episode he was asked by a serving private soldier, 'Please tell me, if your name is on an enemy bullet, will you get it?' Joad replied without hesitation, 'Well, if your name is actually on the bullet I do indeed believe that you will get it. But the more interesting question that occurs to me is, what will happen if it is written on the bullet, "To whomsoever it may apply!"'

My final year in Uganda was spent in a newly-built house on the recently acquired University farm at Kabanyolo, some fifteen miles north-east of Makerere. By that time five extra 'agricultural' staff had been recruited by

Fergus, together with another ten or so 'pure science' staff stationed on Makerere Hill. Willem Boschoff and his family were our closest friends at Kabanyolo, and we had a most rewarding time putting the finishing touches to the new estate. One of the missing features was a set of tennis courts and as this was deemed to be of no academic value, we simply had to set to and make the courts ourselves. We used a mixture of anthill dust and *murram* sand (derived from a red sandstone common in Uganda). We made quite a good job of these two tennis courts and I like to think that, with a new surface rolled down every three or four years, they may still be serviceable and giving pleasure to the Kabanyolo residents.

After six and a half full and exciting years, our sojourn in Uganda came to an end. We had learned much and I had taught a little, but the advantage was definitely in my favour. I had done some useful research, which was read and appreciated in the UK. It is gratifying that it is still referred to today. John Hammond's advice had been very sound. We had travelled widely, seen many interesting sights and got to know many Africans, far better than many expatriates who had lived in the country for more years than we did. We had come to appreciate the generosity of spirit and inherent friendliness of many Africans – ranging from small peasant farmers to professional staff and local chiefs. Politically, Uganda was a happy, stable country, with law and order maintained with a relatively small local police force and three companies of the East African Rifles. We had seen the beginning of the end of the Colonial Empire, but fortunately we left several years before independence and the rise and fall of Idi Amin. Idi was known to us, as he was then serving as a NCO in the East African Rifles, where he was popular with his European officers as he was the Army boxing champion and also one of the few Africans to play rugby football. We spent wonderful family holidays in the Kenyan Highlands, staying on the Pleadon's farm. They were good 'growing' years, with new and exciting developments taking place on the campus and more of the Makerere graduates, like Julius Nyerere, taking up prominent positions of responsibility in their respective countries.

I was extremely happy throughout my time in Africa, but Bunny was less so as the nature of my work meant that she was left alone in very isolated places, the servants having retired to their own quarters at night, two small children to protect and the rumble of native drums going on until very late. With Mau Mau raging in Kenya only a short distance away, she never knew whether the dreaded 'invasion' of Uganda had begun and often slept fitfully in her bed whilst I was working far from home. But the good memories outweighed all others. We made real friends, with whom we have kept up over the intervening years. The Africans were wonderful, proud but kindly people. Their large extended families meant that there were never any problems of orphans or unwanted children — all were taken into the close circle of the family unit. The days were bright and sunny and the rainy seasons were relatively short. There were no long, cold winter nights in Africa! It has been sad to watch from afar the troubles of Idi Amin's reign and the present problems in many other parts of Africa. I have never returned to East Africa — I know that to do so could be painful. I prefer to remember the country as it was in those early post-war days. We left much wiser and more experienced people than when I had signalled from London seven years previously,

Purchase two grass skirts.

CHAPTER 8

Trinidad

'Go West, Young Man'
John Babsome Lane Soule

A FTER SIX YEARS IN Uganda as an Assistant Lecturer and later as a Lecturer, I felt the need to take up other employment, gain promotion, and thus increase my stipend. Raising a growing family on £650 a year was not easy, and it did not seem to me that there were any early prospects of advancement under Fergus Wilson, whose philosophy in such matters seemed to be: 'one small step at a time!' So I began to look around.

Two posts came up at the same time, both at Senior Lectureship level, one at Queensland University, Brisbane, Australia and one at the Imperial College of Tropical Agriculture (ICTA), Trinidad. I applied for both posts simultaneously. The Australian post progressed quite rapidly to a 'near offer', subject only to the approval of the Head of Department, who was absent on a sabbatical year. Nothing could be done, apparently, until the Professor's return. The Trinidad post moved more quickly. I was told my application had been favourably considered, but to finalise matters I would need to meet Sir Frank Engledow, the Professor of Agriculture at Cambridge University, who was Chairman of the Appointments Committee of ICTA. Frank Engledow was also on the board of the Tea Research Institute, Kericho, in Kenya, and as he was making a routine visit there, I was asked to travel over to Kenya to meet him.

I duly drove to Kericho and had a long interview with Frank Engledow. I learned that the Senior Lectureship for which I had applied was vacant because the recent occupant had resigned to take up a more senior post in Florida, USA. It so happened that I knew the man in question, Bobbie Howes, who had been a contemporary of mine at Wye. Being ex-service, he was senior to me in years and I knew that he had made quite a name for himself by building up a highly regarded Department of Animal Production and had published many papers in the animal science field. He had recruited a junior lecturer, Ross Houghton, also an ex-Wye man. Being younger than I, and therefore of a later vintage, I did not know Ross, but the prospect of having a colleague with a Wye background was an added attraction.

The ICTA post entitled the holder to UK leave every two years. The Australian position, on the other hand, necessitated emigrating to Australia 'for life' and any visits to the UK would have to be self-funded. Bunny was not too keen on severing UK connections in this way. We both had elderly parents to think about and our two children's education to plan, and so the dice were weighted heavily in favour of travelling west to Trinidad rather than south to the Antipodes. I accepted the ICTA offer and a few months later we set sail on the SS *Durban Castle* from Mombassa, via South Africa, to the UK, from which, after a four-month holiday, we embarked on the SS *Camito* from Bristol docks, bound for Trinidad.

On our arrival at Port of Spain Harbour ten days later, we were greeted on the dockside by Ross Houghton. We were travelling not only with all our luggage, but also with a recently acquired new car, and Ross pointed out that we would need a Trinidad driving licence before we could move it from the docks. As there was no time like the present to attend to matters of this sort, we were driven to the licence office, which was run by the Police Department. Here it was made clear that the process of getting a licence was a protracted one and would probably take several hours. I conferred with Ross as to how the red tape could be curtailed, as spending our first few hours in Trinidad in a police station was not, in my view, a very worthwhile way of passing the time. Ross thought the matter over and said that the police on duty at the licence office were rather bored with their job and perhaps we could help them to spice it up. We decided on the tactic of placing a bet with them that the whole process could not be completed in less than half an hour. This challenge was enthusiastically accepted, and we were hustled from one office to another with an ever-growing pile of stamped papers until, some twenty minutes later, we emerged clutching two brand new Trinidad driving licences. The bet had cost us about twenty dollars! This incident taught me two very important things about the West Indian's psyche. Firstly, no West Indian can refuse a wager, especially if he is likely to gain from it. Secondly, the way to win friends and influence people in Trinidad was to appeal to their sense of humour. Many times thereafter, employing one or other of these ruses has got us out of what might otherwise have been quite tricky situations.

ICTA was founded in 1923 by the British Government as a centre for agricultural teaching and research. It served not only the West Indies, but the whole of the Colonial Empire, as its postgraduate students were deployed throughout the overseas Agricultural Service. In Trinidad they took a one-year 'Associateship' examination (AICTA), in my time upgraded into a Diploma in Tropical Agriculture (DTA). Most of the crop research was commodity-orientated, and concentrated on coconuts, cocoa, sugar-cane and bananas. All other tropical crops were dealt with by staff members recruited with the required specialist experience. Animal science research, however, was conducted solely within the Agriculture Department, with a small staff of only three people. When I first went to Trinidad, the administration was in the hands of the Colonial Office, who delegated most of the work to a Board of Management under the chairmanship of Sir Frank Engledow. He in turn gave responsibility for the day-to-day running of ICTA to the Principal and his team of Professors. The West Indian students took undergraduate diplomas up to 1961, when degrees of the University of the West Indies were introduced. It was interesting to note that the former diplomas were of degree equivalence, so when the qualification was upgraded, the pass rate remained unchanged even though the first degrees were awarded by London University.

The Agricultural Department was the largest and most important Department in my time, and since 1923 it had been headed by a Professor appointed on the strength of academic merit and relevant experience of tropical agriculture. When I arrived, the Professor of Agriculture was Dr Tom Webster, another ex-Wye man, who had served in Malaysia and Central and East Africa, his most recent posting being in Kenya. Tom Webster was a quietly spoken man, with that most important ingredient, a twinkle in his eyes and a well-developed sense of humour. He was shrewd and highly motivated. He was an excellent teacher and was popular with the postgraduate students, who did not suffer fools gladly. He supported his staff, especially if they were active in research, but all requests for research funding were most carefully examined and if there was any woolly thinking or slack budgeting, it would be quickly exposed. He had a kindly way of saying 'no', which usually invited the staff member in question to 'Think again, with perhaps greater reference to points a, b and c'. He was also a good listener, and one of his ploys when dealing with verbose expositions was to remain silent until the person concerned had either run out of steam or backed himself into a corner of his own making. Tom would then smile broadly and say, 'I think we had better sleep on this one until we get it properly thought through!'

I enjoyed the challenge of teaching both West Indian students for their undergraduate courses and the postgraduates for their DTA. The West Indians came from all the Caribbean territories and it was soon apparent that the cultural differences between the Islands were as wide as the racial differences between the students I had taught at Makerere. I learnt that it was derogatory to refer to West Indians as such. They were Jamaican, Guianese,

Trinidadian, Barbadian and it was important to recognise their different accents. Calling a Trinidadian 'West Indian' is equivalent to calling a Scot a European (or, worse still, an Englishman!). Trinidad, in particular, was a very polyglot community, with people not only of African descent, but Asians, Portuguese, French, Spanish, Chinese and British, together with their various reciprocal crosses. The Carib Indians, the earliest inhabitants of the West Indies, were few in number. Most of them had been wiped out by contracting European diseases or by the rigours of their harsh life in slavery. The few survivors cherished their traditional customs, such as the crowning of the Trinidad Carib Queen whenever the position became vacant.

The postgraduate students varied in number from year to year. In the peak year – 1960 – there were over one hundred, but in my latter years in Trinidad the numbers fell as more of the former colonies became independent. There are now no longer any students being sent to Trinidad for training from the UK, which is very sad. A cadre still exists of British agriculturists who have received the benefit of a formal course of tropical agricultural teaching in the tropics, as distinct from a theoretical course in a British University lecture room. Inevitably, however, it will fall to zero. There are now far too many 'experts' rushing around the tropics with a briefcase full of typed reports, dispensing 'advice' from a lap-top computer with only the vaguest notion as to the realities of rural life at the sharp end. Recipient governments must swallow their pride and accept these 'advisers' with a pretence of respect, as large grants from international agencies hinge upon whether or not the advisers make a positive recommendation to their parent body. I fear that much of this 'advice' will be wide of the mark, and many millions of dollars of aid has been, and will be, misspent as a result.

During my six and a half years in Trinidad, over 350 postgraduate students passed through the DTA course. Many of them are still alive and some still active, but the number surviving is decreasing each year. All these students had attended an intensive course of lectures in relevant subjects. In addition, each one of them had undertaken a research project and written up the results in the form of a dissertation. I encouraged the students to regard the project not just as an academic exercise, but to consider the work as part of a programme of agricultural research which, hopefully, would eventually be published. Most students were so busy getting involved in day-to-day problems in their first postings after leaving Trinidad that they did not take this advice, but I know of six papers based on these postgraduate projects which were eventually written up and published (17, 18, 19, 23, 29, 30).

Some of the best students knew what was expected of them before they arrived in Trinidad. One outstanding student, Dennis Osbourn, wrote to me whilst at Cambridge, spelling out the research project he would like to conduct. He had done his homework and knew of my interest in the growth and development of farm animals and had read my papers describing my work with poultry and the East African dwarf goat in Uganda. He sent me a very detailed research proposal describing an experiment on compensatory growth he wished to carry out on beef cattle. Could I supply 48 animals accordingly and have them ready for his arrival? The work was of interest to me and I wished to encourage Dennis, so I wrote back and said I liked his proposals but could he scale the project down to 48 poultry instead of 48 beef steers, as our budget would not run to bovines. And so it was. A year later Dennis and I published two major papers on compensatory growth, concluding the long string of experiments on farm animals in the 'Hammond school' tradition (17, 18). These papers are still quite widely quoted and Dennis' initiative was well rewarded.

Dennis obtained his DTA with distinction and I had high hopes of a senior position for him at an important Animal Research Station. The Colonial Office had other ideas, and just to show that they were in charge, Dennis was sent to Tanzania to take charge of a pyrethrum breeding scheme. 'Horses for courses' was not a concept the Colonial Office understood and I can imagine a junior official in Whitehall saying, with a broad grin on his face, 'We'll show them who is in charge of this outfit!', though I am quite sure Dennis did an excellent job with pyrethrum, he hankered for an animal production post and when eventually Ross Houghton left me to join the Buxted Chicken Company in the UK, I brought this vacant post to Dennis' notice. He duly applied and got the job. Dennis made an excellent colleague and I was very sad when he eventually left me, on promotion, having been head-hunted to go to the Grassland Research Institute at Hurley.

Another of my close colleagues was Trevor Chapman. He had obtained his DTA before I arrived in Trinidad, but I recruited him as the senior crop husbandryman on the staff soon after I got the Chair following Tom Webster's departure. I had been informed by many colleagues that he was anxious to come back to Trinidad, and when a suitable vacancy came up, I drew the post to his attention. We duly met at his interview in London and I was most impressed. He was offered, and accepted, the job and was a tower of strength to me on the planning side of our work. He was an independent thinker and we often had amicable disagreements concerning the best policy for the Faculty, but we remained firm friends throughout our joint periods of office. On several occasions we took our family holidays together in Tobago, at one time staying as joint guests of Tony Bishop on his coconut plantation near Scarborough. Trevor left Trinidad to join the staff of Shell Research at

Woodstock, Kent, as an agronomist, eventually becoming Director of the Station. He is now retired in Cornwall and lives in a lovely house only a few miles from Dennis Osbourn, which allows the two families to see a lot of each other. Trevor and Janet Chapman had a son, Roger, who was a pupil at the Senior Staff School when Bunny was Headmistress (see chapter 19) and it was clear to everyone that Roger was a natural ball-game player. If you threw a ball to him, he could catch it with ease; if you gave him a bat or a racquet, he could hit a ball for six without thinking and it was clear that a great future lay ahead of him. In fact, he took up golf and is now a 'pro', playing in most of the international 'Open' championships around the world.

Our Trinidad house was situated on the College farm and our immediate neighbours were Peter Rosher, the Senior Lecturer in Agricultural Engineering, and his delightful family. Peter was a specialist in the application of pesticides to tropical crops, including the new generation of weed killers. Peter developed a reputation for conducting field trials into new pesticides and this brought him to the notice of Shell Chemicals. He was subsequently offered a post with Shell at their research station at Woodstock in Kent. Peter was very keen on Trinidadian music and had a vast collection of the works of the leading calypso and steel-band musicians. He soon had us 'hooked' on this aspect of local island culture.

When Peter left Trinidad to take up the job with Shell, his post was taken by Lewis Campbell, so Lewis and his large young family duly became our new neighbours on the farm. Lewis was one of the first West Indians to obtain a senior staff appointment at the University, and he played a most useful role in helping to smooth the process of West-Indianisation and enable it to progress steadily and without friction. When I left, there were about a dozen West Indians occupying senior staff appointments. I look back with great satisfaction on the fact that all these changes were introduced without any problems or racial difficulties, a comment that cannot be made about many of the other ex-colonial universities achieving independence at that time.

I was very anxious to enhance the status of our West Indian students, who, until my period in Trinidad, could only obtain diplomas in agriculture and not full University degrees. One of my outstanding diploma students was Tom Henderson, who came top of his class and took away a clutch of prizes in a variety of subjects. Having a vacancy for the position of Manager of the University Farms, I offered Tom the job. He accepted it and more than fulfilled all my expectations for him. Strangely, he had more difficulty in making his mark with the local West Indian farm staff than he did with his expatriate academic colleagues. One might have expected that there would

*26. The author and Mr Almeida, Senior Technician, checking a sisal plant on the
College Farm, Trinidad, 1958.*

have been genuine pleasure in seeing a 'local lad make good', but the farm
staff took some time to accept him as their 'boss'. Tom subsequently went to
America to take a post-graduate degree, but unfortunately he and his family
came under the influence of the 'Black Power Movement', which was a force
to be reckoned with in the USA at that time. I was extremely sorry that the
goodwill which had embraced him in his Trinidad post was not matched by
his reception in the USA. The early 1960s were difficult days in America,
and Tom had chosen to study in one of the southern US Universities, where
racial problems were at their worst

The postgraduate students were able to choose a research project close to
their own field and interests, and this resulted in a large proportion opting for
an animal production subject. In 1959 I had twenty-two students desirous of
working in an animal-related field and it was difficult to find good projects
for such a large number. One method of tackling this problem was to deploy
students in a row of eight for the purpose of conducting studies on animal
behaviour. This work involved each student doing a three-hour shift,
observing a herd of cows for a continuous period of 72 hours and recording

their behaviour in terms of time spent grazing, ruminating, walking, lying idle, drinking or asleep. We found that one could alter the behaviour pattern by offering better quality pastures by night and poorer ones by day, and in this way we could shift the animal activity so that more grazing took place during the cool nights and less during the hot days. The average times spent in these activities were roughly about eight hours grazing, eight hours ruminating, and eight hours walking, lying idle, drinking or asleep.

One interesting point, which the teams of students found fascinating, was that less than ten minutes in every 24 hours was spent sleeping – during an idle period between midnight and 3 a.m. Some cows slept for only about 5 minutes, and after this very short sleeping period, the cows immediately returned to grazing. The students were all interested in observing this fact for themselves, as they found it difficult to believe. In consequence, there was always a good number of volunteers for the three-hour stint between midnight and 3 a.m. Much of this work was subsequently written up and published (22, 25).

These cattle behaviour trials became quite well known on the campus and caused me to write a poem, in a light vein, describing how we went about our task. The final verse ran as follows:

> The question posed, you will agree
> Is 'Are the cows observing me?
> Are my movements round the field
> Discussed, debated, cow-revealed?'
> For only Bovines can decide
> If Man is fully occupied!

The students worked hard and played hard in Trinidad. The West Indians were able to field a most excellent cricket team, easily up to county standard, although their opportunities to play against matched teams in faraway places were perforce somewhat limited. Occasionally the British postgraduates contributed to success on the cricket field, but their main sport was rugby, and they acquitted themselves well playing in 'mini-international' games against other West Indian territories, especially Barbados and British Guiana (later Guyana), where their opponents were mainly expatriate managers of large sugar, banana, coconut and cocoa estates.

The tradition in the Colonial Office in those days was that Agricultural Officers were not expected to get married until they had completed at least one tour of duty in their overseas posting. Marriage before the postgraduate year in Trinidad was frowned upon, as it was thought to undermine the 'mobility' of the officers to go to remote postings, where living conditions

27. The main building, the Imperial College of Tropical Agriculture, Trinidad, 1960.

were primitive and deemed to be unsuitable for wives and children. I thought this policy was crazy. Thus, whenever I had the chance, I visited the postgraduates at Cambridge during the year before they sailed to Trinidad and told them frankly that they were not supposed to bring wives with them to the West Indies, since there was no suitable accommodation for married students on the ICTA campus. However, rented accommodation was not difficult to find and if the wife had some formal qualifications, there was no difficulty in getting a permit to work. Many students took the hint (including Dennis Osbourn) and sailed out to Trinidad with a wife in tow. They were much happier than they would have been leading a bachelor existence, and the wives benefited by having some useful tropical experience in the West Indies before being thrown in at the deep end when posted to some remote part of the Colonial Empire. Sometimes there were a few Australian or New Zealand postgraduates in the group and these proved to be even more resourceful. They thought nothing of buying a plot of land, a few tons of bricks and a pile of timber, in order to build their own small houses near St. Augustine, where the College was located. On leaving Trinidad, they invariably managed to sell their self-built property at a fair profit.

On the other hand, I had noticed that some of the ICTA staff recruited from the UK who were allowed to bring their wives with them had great difficulty settling into their new environment. The wives had little or no knowledge or experience of life in the tropics, and some of them were so unhappy with the very different social life in Trinidad that they were soon anxious to return home. This meant either a very short period of service for the staff member concerned or else the break-up of a marriage. I decided that we needed a much higher 'success rate' with adjustment to life in Trinidad, so I argued that we should be seeing the wives as well as their husbands, when interviewing candidates for a lecturing post. This, I was told, was definitely not on. We were appointing staff to do a job, not staff wives to enjoy a happy social life in Trinidad. However, I persevered and arranged that, before or after the formal interview, I would entertain the successful candidate and his wife to lunch or dinner, so as to form an opinion about their joint suitability. Doubtless this was, and still is, very 'politically incorrect' but there were no more disastrously short periods of service after I introduced the new procedure.

The second instance of Colonial bureaucracy concerned the running of an experiment on the effects of different stocking rates of fattening cattle on liveweight gain (14, 16, 19). The experiments were carried out on the Government Stock Farm, as ICTA had no suitable facilities. This farm was about ten miles distant by road but only two miles on horseback and as I was then the proud owner of a stallion, recently retired from running in the Trinidad races, I argued that it would be less expensive if I was allowed to travel on four legs instead of on four wheels. I was fully entitled to payment for the twenty-mile round trip in my car but not, according to regulations, to the four-mile round trip on horseback. Correspondence flew backwards and forwards from Trinidad to Whitehall and I was getting nowhere, until I met up with a former Administrative Officer who had served as a District Commissioner (DC) in Nigeria. Northern Nigeria at that time had very few serviceable roads, so DCs were forced to get round their District on horseback. I quoted this precedent to the Colonial Office and won my case, 'provided I charged the travel in the form of an 'oat allowance', at the same rate per mile as applied to a private car'. My horse was never actually fed on oats but he earned his keep in other ways, and I saved myself a lot of time in unnecessary travel as I watched the progress of the experiment. Apart from this, the daily ride on horseback over to the stock farm was most enjoyable.

The Principal of ICTA for much of my time in Trinidad was Dr Geoffrey Herklots. He was a small, slight, somewhat serious individual, much more comfortable in shorts and a bush shirt than in formal attire. A botanist by training, he had worked before the War in the Far East and was captured by the Japanese at the fall of Singapore. Geoffrey was interned during the War with his wife, and many internees owe their lives to him as, with his knowledge of botany, he was able to augment the meagre camp rations with edible wild plants growing around the compound. In addition to being a good taxonomic botanist, he was also an ornithologist and he spent much of his spare time in Trinidad in the middle of the rain forest of the Northern Range of mountains, searching for birds. He was not interested in bird behaviour or bird ecology; his only objective was shooting the birds and bringing them back to his home to paint. His coverage of the local birds was almost complete and his book on the *Birds of Trinidad and Tobago* is regarded as a classic.

Knowing of his interest in the birds of the forest, the students asked him to give a talk on the subject and I went to his well-attended lecture. For me the talk was disappointing. It consisted of a long list of birds together with a few anecdotes about the difficulty in locating and shooting some of the rarest species. At the end of the lecture Geoffrey asked for questions but there was only one: 'What sort of a rifle do you use to shoot the birds, and when you have shot and painted them, do you ever eat them?' (The answer to the first question was 'a .22 rifle' and to the second question 'no, never!'). Unfortunately this lecture stifled my interest in birds for many years. I am now a keen, but very amateur, bird watcher and much regret that I missed a golden opportunity to study the birds in Trinidad.

Geoffrey also collected orchids. Indeed he combined orchid collecting with bird hunting whenever possible, and he had a large collection of indigenous and imported orchids. When he retired, his collection was handed over intact to some form of Trust and at the time I left Trinidad in 1964, it was still well maintained as the *Herklots Orchid Collection*.

On one occasion Princess Margaret paid an official visit to ICTA, spending most of the day looking round the departments and meeting staff and students. A small band of children from the ICTA staff school was assembled by their head mistress to greet the Princess. Unfortunately they sought cover from the hot sun under a *Cassia* tree, which, unknown to the staff, contained a large wasps' nest. The flag waving upset the wasps, who were distracted from going about their normal business and attacked the children. They fled in all directions, beating a hasty retreat, just as the Princess arrived at the campus.

The visit ended with a luncheon party in the Principal's residence and Mrs Herklots requested me to produce a quart of cream from the University dairy herd. Now we did not normally separate the milk into cream and skim milk. I knew we had an old Lister milk separator and I duly had this cleaned up and proceeded to demonstrate to the staff how the machine operated. It was an old-fashioned model, the separator being worked by means of a wooden crank handle, which drove a flywheel attached to the churn, which separated out the cream by centrifugal force. Just as I was in the process of getting it turning at high speed, the old wooden handle split and a large splinter flew off, badly cutting my thumb in the process. I realised that if I stopped the machine I would never get it going again, so I rapidly bound up my bleeding thumb with my handkerchief whilst keeping the mechanism moving. The result was that a little of my blood got into the cream, turning it a very pale pink. However, as there was no prospect of repeating the operation with a fresh lot of milk, I duly delivered the cream as requested, and got my injured thumb sown up in the College dispensary. The cream, I am glad to say, went down very well, fresh cream being something of a novelty in Trinidad!

The luncheon was not without its amusing moments. The Herklots rarely drank and normally no alcoholic drinks were served to their guests. In honour of Princess Margaret, several bottles of sherry were purchased and all the guests were offered a glass before the meal. Princess Margaret, not being a sherry drinker, declined. A Lady in Waiting, noting this, went up to her and said, 'Ma'am, would you like your usual gin and tonic?' 'Yes please!' she said, so the Lady in Waiting went to Geoffrey Herklots and told him that the Princess would prefer a gin. There was no gin in the house, but Geoffrey knew that the rather eccentric Professor who lived next door, drank rather a lot of gin, preferring to buy the local brew, which was half the price of imported Gordon's. A servant was despatched to borrow a bottle of the beverage and returned bearing a full bottle of gin together with a large tumbler. He duly presented HRH with the unopened bottle and one large glass. As there was no tonic in the house, Princess Margaret drank a few fingers of neat gin before proceeding to the dining room, whilst the rest of us had to make do with medium-dry sherry!

On arrival at the luncheon table, Princess Margaret waited until the company were assembled and then graciously sat down, indicating that the other guests could follow her example. Geoffrey was most concerned at this premature start to the meal. He rapped the table and exclaimed 'We usually start with grace before the meal!' HRH rose hurriedly to her feet, grace was duly pronounced and the luncheon commenced. At the end of the meal coffee was served. Now the Herklots had only a dozen coffee cups and saucers and as there were 18 guests, he arranged that the collection would be increased by the addition of some borrowed cups and saucers. These, the servants were instructed, would serve for himself, his wife and some other staff who had been forewarned to use the odd china. Unfortunately the plans went awry, and the servant dispensing coffee offered Princess Margaret the whole trayload of cups and saucers, with the borrowed pairs unfortunately perched on the top. The Princess took one odd cup and one of the saucers, but seeing she had an unmatched pair, put back the cup and had to search around to match the saucer, unfortunately ending up with one that was cracked. That evening, as was the custom, HRH returned the hospitality by inviting the Principal and his luncheon guests to a reception at the Governor General's residence in Port of Spain. In talking to one of the Ladies in Waiting, Geoffrey enquired whether the Princess had enjoyed her lunch at ICTA. 'Oh Yes', was the reply, 'I am sure she did. Indeed, she was talking about nothing else all the way back to Government House!'

When we arrived in Trinidad, the country was a colony. After three years the British Government endeavoured to form the numerous West Indian Islands into a Federation, which, in terms of administrative convenience was clearly a step in the right direction. So in 1959 Bunny and I found ourselves on the docks of Port of Spain welcoming the first (and only) Governor General of the Federation of the West Indies, Lord Hailes. It was a most splendid affair; flags flying, bands playing, welcoming speeches – all the pomp and ceremony for which the British Empire was so rightly renowned. Only two years later the West Indian Federation had broken up and Bunny and I were once again on the dockside at Port of Spain bidding farewell to Lord and Lady Hailes as they took their leave. There cannot be very many political institutions which have risen and fallen quite so rapidly.

Not only the Trinidad political constitution but also the academic administration of the College changed drastically during our six-and-a-half years. When we arrived, my appointment was with the Imperial College of Tropical Agriculture. Three years later ICTA was absorbed into the University College of the West Indies, based in Jamaica, and our undergraduate courses were upgraded to external degrees of the University of London. The scheduling of new degree courses for the students was a task with which I was familiar, as it was a parallel exercise to the one which I had been party to in Uganda. However, the West Indians were not content with being an appendage of London University. After another two years the University obtained its own charter and became an independent University. The syllabus had to be re-written yet again, in order to give it a greater West Indian content. The focus was changing from a primarily postgraduate institution, ICTA, to a University majoring on courses tailor-made for the specific needs of the agricultural problems of the Caribbean. As far as was possible, I made it my business to focus on principles based on a solid foundation of science and economics. For it was clear that while, over time, the problems of the day would change, the basic principles would still be

28. *Our son, David, bidding farewell to Lord Hailes, former Governor
General, on the occasion of his departure at the end of the short-lived
West Indian Federation, 1962. Bunny, far right; the author, behind David.*

relevant, however much land-use and cropping patterns altered over the years
(24).

In 1960 Tom Webster announced that he was resigning from the Chair of
Agriculture to take up the position of Director of Rubber Research in
Malaysia. He suggested that I should put my name forward as a candidate for
the Chair, which I did. I was duly interviewed and was successful in getting
this interesting post at the exciting time when both Trinidad and the
University were becoming independent. There was much speculation
amongst the staff that political considerations would require the Chair to be
filled by a West Indian and indeed several did apply. By then we had already
recruited a lot of West Indians onto the teaching staff and so perhaps the

pressure to 'West Indianise' the Chair was not as acute as it might otherwise
have been. I was, however, well aware that I was fortunate in getting the post.
That said, I never once experienced any feeling of racial hostility against me
and I enjoyed very happy and cordial relationships with my West Indian
colleagues. I was the ninth (and last) British expatriate to hold the Chair of
Agriculture in Trinidad (The list of the other occupants of the chair is given
in Table 1).

Table 1
Professors of Tropical Agriculture in Trinidad

1923-1927	Professor J.S. Dash, BSA
1927-1937	Professor R.C. Wood, MA, Dip Agric.
1938-1942	Professor E. Harrison, CMG, B.Sc., MSA, NDA
1942-1946	Professor D.D. Paterson B.Sc.(Agric), B.Sc.(For)
1946-1949	Professor A. de K. Frampton, CDA
1950-1951	Acting Professor C.W. Lynn, MBE, CDA. AICTA
1952-1955	Professor A.B. Killick, CMG, B.Sc., AICTA
1956-1960	Professor C.C. Webster, B.Sc., Ph.D., AICTA
1961-1964	Professor P.N. Wilson, B.Sc.(Agric), M.Sc., Ph.D
1964- ?	Professor P. Mahadevan, B.Sc., M.Sc., Ph.D (from Sri Lanka)

I have never been an active trade unionist. However, when ICTA became
part of the University of the West Indies, it was deemed appropriate that the
staff should be encouraged to join the West Indian branch of the Association
of University Teachers (AUT). I duly joined and much to my surprise, I was
elected the first Chairman of the Trinidad branch, and as such had to find
something useful to do. I discovered that other overseas AUT branches had
successfully persuaded their University Authorities to pay all, or part of, the
staff contributions to the BUPA health insurance scheme, which provided
free medical attention for those that required it. I therefore asked the
University of the West Indies to fund subscriptions to BUPA. To my surprise,
our request was well received, and when I left Trinidad, not only did staff
enjoy these medical insurance benefits, but the fact that we had gained this
concession meant that we secured a very high-percentage enrolment of
Trinidad-based staff in the AUT.

Being now in a position of trade union respectability, I ran into an
interesting examples of trade union problems at a lower level. After ICTA
had joined the University of the West Indies, the local administration was no
longer in charge of the 'pay and ration' of our staff, which had to be agreed
on a University-wide basis. Unfortunately the timing was against us, since we

joined the University when negotiations about the terms and conditions of our local manual staff were almost, but not quite, complete. The merger with the University meant there were long delays in settling the dispute. Not unnaturally, the lower-paid manual staff in Trinidad were getting agitated about their falling standard of living. Philip Sherlock, our newly arrived West Indian Principal, failed to grasp the significance of the problem and did not expedite the pay settlement. The inevitable consequence was that all the manual staff, including the farm staff, went on strike. It was bad enough that the students did not get fed and the premises were left uncleaned, but it could have been a major catastrophe for the milking herd and for other livestock on the University farm.

In those days we lived on the farm and a few days before the strike started, a deputation of the farm staff came to my house to see me. 'Boss,' they said, 'we have a problem. Our union says we must come out on strike and if we don't, our wives and children will be beaten up.' I agreed this was serious and asked them what they thought we should do. 'Boss,' they said, 'we've got it all worked out. There is another large dairy herd at Point a Pierre attached to the oil field, and their farm staff are not members of our union. So if we move the herd down there, the cattle will be milked and looked after for as long as the strike lasts.' I congratulated them on their helpful advice and that night a fleet of cattle trucks transported all the College milking cows some 50 miles to the south.

The farm staff were all very apologetic about the strike, which was not of their making. However, because of the threats to their families, they could not indulge in strike-breaking. I agreed. 'Boss,' they said, 'we've a favour to ask. We have been told we must picket the main farm drive and not let anyone come in or out.' 'That could be difficult,' I said, since I was not on strike, 'What about me?' 'Don't worry, boss,' they said, 'we will let you and your family in and out whenever you want, but we must have a picket and we must have a big banner saying that we are on official strike!' 'OK,' I said, 'but I think you should have a banner for me too.' 'What should it say?' they asked. 'More money for the Prof.', I replied, with my tongue in my cheek. Imagine my surprise next time I returned to the farm to find displayed a beautiful, brightly coloured banner with 'More money for the Prof' duly written in large letters. As I mentioned previously, if you could make Trinidadians laugh or smile, you never had any trouble with them. They are a marvellous people, generous in spirit and with a highly developed sense of humour.

Three other little incidents illustrate the Trinidadian sense of fun. I was keen on amateur dramatics, and one of the productions I played in was

Murder Mistaken, in which I had to eliminate one of the characters with a heavy poker. On the night of the first performance I was delayed in Port of Spain by a major traffic jam, and it was clear that, without some help from a risk-taking taxi cab driver, the play would start somewhat late. I hired a cab which had not yet got a full payload (taxis at that time went on picking up passengers until they were full) and asked the driver to get a move on as, I explained, 'I have to murder a woman in twenty minutes'. With a broad smile he accepted the challenge and as we passed groups of unhappy potential passengers along the way, he leant out and shouted, 'No room for you, man, I have to help this gent murder someone in twenty minutes!' We made it on time and the curtain went up only two minutes late.

On another occasion I came across a heated argument between two Trinidadians in the main street of the capital city. Each was aggressively berating the other, with taut bodies and heads and shoulders jutting out in menacing fashion. A local passer by, inconvenienced by this confrontation, shouted at them, 'Pull in yer lip man, an' let me pass!' Many a humorous word turneth away wrath in Trinidad!

The third incident is perhaps even more revealing of the importance of humour to the Trinidadian, coupled to his sense of 'fair play'. We were very friendly with a Chinese family, the Lee Lums, who had come to Trinidad from Hong Kong. Edwin Lee Lum had been a millionaire no less than three times. He had created a large business empire in Hong Kong, which had run on to the rocks. He had then emigrated to Trinidad, where he proceeded to found another business empire. This, apparently, ran into difficulties during the Second World War. After the War he started up for a third time and it was during this period that I got to know him well. Part of this Lee Lum 'Third Empire' was a venture into deep-sea trawling in the Caribbean for tuna fish. Up to then, the local fleet had been confined to shallow off-shore fishing, but the Caribbean waters are deep and bigger creatures were abundant at greater depths, since hitherto no fishing gear had been able to reach them. Edwin's new venture did well and his deep-sea trawlers came back to harbour with holds full of valuable tuna. Now the dockers in Trinidad were a powerful group, very similar to the stevedores of America, and they saw a way to make an 'easy killing' out of these profitable tuna fish landings. They demanded the right to unload the ships, stating that the minimum 'crew' for each crane used for unloading the catch was nine people – one to waggle his finger at the crane driver, four to place the hooks onto each corner of the fish containers on board the ships and another four to unhook them when they had been swung onto land. Not only was this expensive, but as the dockers were being paid partly on a time-basis, the tuna began to thaw out on the quay and fish quality was becoming a major problem.

Edwin decided to call the dockers' bluff. 'It doesn't require four men to put the hooks on or off,' he said, 'One man can do the job!'. 'You don't know what you're talking about,' was the reply. 'It takes four people – eight in all.' Edwin challenged the dockers and said, 'If I (an old man of 80) can do the work on my own, then surely one hefty docker could easily do the job!'. As I have already indicated, a Trinidadian loves a gamble and so they bet Edwin that if he could do the job for a full shift, they would withdraw their demand; but if he failed, he would pay them $1000. The bet was accepted and on the appointed day Edwin turned up, stripped down to a pair of dirty shorts. He duly worked in the heat of the tropical sun, cheered on by a large crowd of rowdy onlookers. When Edwin won the bet, however, the dockers warmed to him. They hoisted him onto their shoulders and carried him off to the nearest rum shop for a marathon drinking session. A week later they made him an Honorary Life Member of the Dockers Union and he became something of a local hero. If that anecdote does not demonstrate a sense of humour and fair play, I don't know what does!

The University Faculty of Agriculture had an Advisory Council made up of prominent Trinidadian agriculturists and other public figures. One such person was Sir Hugh Wooding, the Chief Justice. He was a member of a much respected Trinidad family of African descent and we became very friendly. He was a strong supporter of our work, but in spite of his loyalty, I detected there was something at the back of his mind which made him uneasy about the former ICTA, and when I had got to know him well, I asked him what it was. 'It all goes back a very long way,' he said. 'When I was a young man with a newly acquired law degree, I joined one of the main law firms in Port of Spain. My very first assignment was to go to ICTA to get the Principal's signature on some papers relating to the transfer of some title deeds. I arrived at the College dressed in my best lawyer's pin-stripe suit and knocked on the door of the Principal's study. I was summoned in but as soon as the Principal saw my black face, he barked at me, 'What the devil are you doing here? If it's a manual job you are wanting, go and see the Clerk of Works.' I have never forgotten that incident, although it happened long ago, for it brought home to me that there were some people who would never accept a West Indian as an equal.' I was very touched and left speechless. How many other similar cases could have been recounted, I wondered, by other West Indians who were insulted as they began to practice their professions in the 1920s and 1930s?

In one of his books Eric Williams, former Prime Minister of the Trinidad

Parliament, quotes a similar incident. Eric won the top island scholarship to Cambridge, where he read history, ending up with a double first in the tripos. At Easter it was the practice of his history don to take a group of tutees to his country residence for a week of intensive tuition and revision. Eric was looking forward to this experience, which, he had learnt, was of great value in achieving a good degree. A few weeks before the arranged visit, Eric was called to his tutor's study and told, with regret, that it would not be possible for him to join the tutorial as 'My family would be a little embarrassed at entertaining a black West Indian.' This hurtful comment left a lasting mark on Eric Williams and it was clear to me that, beneath the surface, it affected many of his relationships with expatriates even when he was the boss and they were his subordinates. His political slogan before he came to power was *Massa day done* (Master's day is over), which was a chant originally used by the African slaves working in the West Indian plantations when they returned to their quarters after toiling all day in the cane fields.

Another member of the Agriculture Faculty Advisory Committee was Geoffrey Stolmeyer. Geoffrey was a member of the plantocracy; he owned a large cocoa estate which had been in his family for many generations. He was one of the most well known and respected West Indians of his generation, as he successfully captained the West Indies Cricket X1 in the early post-war years. When I knew him, he was the Board Chairman of the Trinidad and Tobago Agricultural Credit Bank and he invited me to join him on the Board. Our job was to make loans to deserving West Indian farmers so as to enable them to develop their holdings into more viable units. Each month we duly considered applications and tried our best to make appropriate awards. The problem was that we had only a few staff to monitor the loan, with the result that few if any questions were asked about how the money was actually spent. During the few spot checks that were made, it was clear that the money was not being used for agricultural improvement. Instead much of the money was being utilised to purchase large second-hand cars to start up small taxi businesses. Occasionally, however, the Credit Bank's money was used for the purpose originally intended and such success stories, although rare, gave Geoffrey a lot of pleasure.

Geoffrey was also, for many years, President of the Trinidad and Tobago Agricultural Society. Originally formed as a vehicle for the plantocracy to discuss matters affecting the world trade in the chief export crops, during my time it became an important venue for agricultural lectures. I was asked to address the Society on many occasions. The Society took these lectures very seriously; they were always well attended and published in the Society's journal (12, 13, 20, 21, 26, 27, 34, 35) and I was greatly honoured when the

Society eventually made me an Honorary Life Member on my leaving Trinidad in 1964. I was one of only four such life members, Geoffrey Stolmeyer being one of the other three. After we left Trinidad, Geoffrey was made a Senator in the Trinidad Parliament. He was still revered and remembered as one of Trinidad's most outstanding cricket all-rounders and it was with great sadness that ten years later we heard the news of his tragic death, shot and mortally wounded by an intruder who had broken into his plantation house. It was presumed that Geoffrey was not recognised by the armed burglar, who was high on drugs at the time. His death was sadly mourned not only in Trinidad but throughout the West Indies – indeed over the length and breadth of the cricketing world.

Geoffrey Stolmeyer's large hillside estates were planted to coffee, cocoa and citrus, but the most widespread estates in the island were the sugar cane plantations. Some of these sugar estates were planted to grass on which grazed large herds of water buffalo. These had been imported from India for pulling sugar cane carts, when indentured Indian labour replaced the African slaves after the emancipation of slavery. The water buffalo appeared to be serving no good purpose and I was intrigued as to why this potential commodity was not exploited in any way. I was told that the mule and the small tractor had replaced water buffaloes as a source of tractive power and since the buffaloes were not from a milk breed, there was nothing that could be done with them.

I enquired why they were not used for meat, since buffalo beef has good eating qualities and was much prized in many Far Eastern countries. Receiving evasive answers, I offered to conduct some trials to check out the prospects of raising water buffaloes purely for meat purposes. I found, as I expected, that they could grow to maturity in about two years and I also found that their meat quality was excellent. The stumbling block was consumer acceptability of water buffalo beef. It was said that if discriminating consumers knew they were being offered water buffalo, they would reject it, as it was widely thought that such meat would be tough and inedible.

This offered a challenge which I was anxious to take up. With the support of my colleague, Ross Houghton, I reared a group of water buffalo on good quality pastures and had them slaughtered in the best abattoir in Port of Spain. I organised a prestigious dinner at the Hilton Hotel, to which all the good and the great of Trinidad were invited, including the Chief Minister, Eric Williams. 250 diners duly turned up to a free banquet, the only catch being that they were to be offered three different meats for their main course, A B and C, which they were required to sample and comment upon by means of a questionnaire. Unknown to the diners, these three meats were top quality fresh water buffalo beef, imported Aberdeen Angus frozen beef from the UK and local beef from indigenous beef herds. I must confess I cheated just a little to make my point. The buffalo beef was of the very finest quality and was from animals hung for exactly the right maturation period. The Aberdeen Angus beef was deep frozen, as it had been imported by ship. The local beef was bought from the local wholesale market and reflected not the very best, but the average, quality of meat from crossbred cattle, which was the usual source of West Indian beef.

When the simple questionnaires had been duly counted, the top choice was the water buffalo steaks, which were regarded as juicy, tender, tasty and relatively fat free. The second choice was the imported frozen beef. The local beef came bottom of the poll, although those who mainly dined on such beef found nothing unusual about it. All three meats were superbly cooked by the Hilton chefs.

So we proved to the satisfaction of the top people of Trinidad that water buffalo meat was not only edible but of the highest quality – in fact, the very best! The word soon got round, and some very enterprising friends of mine started herds of water buffaloes and exported both beef and breeding stock to Venezuela and other parts of South America. A senior local veterinarian, Steve Bennett, was the leader of this entrepreneurial group. They named the product 'Calypsolo' beef, and it commanded very high prices. Steve was the most successful and he made good money out of this trade. I published a paper describing the results of our 'Palatability Trial Dinner' in the Agriculture Society's journal (26).

At that time many of the large sugar plantations in Trinidad were managed by expatriate staff. They worked very hard and were paid rather poorly. They were not expected to make a career on the plantations unless they were fortunate enough to be promoted to a top position in the company. One of these expatriate managers was John Meakin, and we got to know the Meakin family very well and have kept up with them ever since. John started farming on a modest scale on his return to the UK and, like so many others, found the going very tough. Farm incomes were dropping and anyone who had borrowed heavily to start up an agricultural enterprise often found that they were working to support the farm, instead of the farm working to support them. I dwell upon this subject more fully in chapter 14.

Many other friends from outside the University were made as a result of their children attending the Senior Staff School, which was attached to the campus and catered for other fee-paying students as well as for the children of members of staff. One of these families was the Taylors. Ken Taylor was employed by the Trinidad Electricity Board and a couple of the Taylor children were pupils at the school. Ken subsequently returned to the UK and became a consultant in mains electricity distribution. They are now settled in the west country and we still visit them when we can. Our respective children have now all married and many have children of their own. It is interesting for Bunny to see the ways in which the lives of children she taught at school in Trinidad have developed over the years.

During my period in Trinidad I was invited by various governments to make consultancy visits to various places in and around the Caribbean. My longest such visit took several months, when I was a member of a British government mission to the Republic of Bolivia. I was the animal scientist on the mission and we arrived in La Paz speaking no Spanish. However, we left many weeks later able to read and write Spanish tolerably well, and we even had the audacity to give an oral report of our conclusions in Spanish to the Bolivian President and his Minister of Agriculture! Staying in a community where no one speaks a word of English is a powerful incentive to learn the local language and it is surprising how much one can pick up by 'the direct method'. I much regret that my linguistic abilities are poor and one of the failures of my school days was that the introduction to foreign languages was so bad that one left school having learnt French or German parrot fashion, without acquiring any conversational fluency. In Bolivia we learnt how to make ourselves understood, but the grammar and the pronunciation must have left a lot to be desired.

Our weeks in Bolivia coincided with some unrest in some of the mining areas, and the police and troops were drafted in to quell potential riots. One of the places we were required to visit was Cochambamba, and the town was temporarily sealed off in an effort to prevent the trouble-makers from fermenting violence. Our visit was an official one and we had letters of introduction signed by the President of Bolivia, requiring all and sundry to give us freedom of movement and any assistance that we might require. We were accompanied by a high-ranking civil servant from the Ministry of Foreign Affairs, whose job was to help us whenever we were in any difficulty. On arriving at Cochambamba, he requested us to let him have our passports and the official Letter of Introduction and quite a sizeable sum of money, as he had to go to the police check point and arrange for us to enter the town for the night. We were happy about the passports and the Letter of Introduction, but demanded to know what the money was for. 'To bribe the police,' was the answer. We remonstrated that with the Letter of Introduction, the bribes were not only illegal but inappropriate. 'But if I do not pay the bribes, they will not let you in,' our guide explained. We argued that we had no spare cash to pay large bribes. 'Don't worry,' he replied, 'You can put the bribes down on your expense account and no questions will be asked, 'When in Rome do as Rome does,' we thought, and duly handed over our cash. As promised, in a couple of weeks' time we submitted our expense claims and were duly paid every penny, including the bribes.

Another very memorable trip was to British Honduras, now called Belize, to report on the damage caused by Hurricane Hattie and to recommend to the British Government the relief measures needed to rejuvenate the agricultural industry. It was pitiful to see the damage done to the city of Belize and to the outer coral-reef islands, the Cays, and to learn that more people were drowned by the tidal wave which followed in the wake of the hurricane than were killed during the storm. Hurricanes are a regular occurrence in the Caribbean, but most follow a west-north-west path, damaging any islands that they pass over as they head towards the North American East coast. Hurricane Hattie, for reasons unknown, started its journey in the usual fashion but then veered west and suddenly turned south, catching British Honduras completely unawares. I am not sure whether our mission was of much help to the farmers of British Honduras. Time is the great healer of an agriculture devastated by a hurricane, and even a week after the storm struck, the new green shoots were showing through the devastation. Forests were decimated and would obviously take a long time to recover, but banana plantations would regenerate the next year and the more sturdy citrus bushes, although they lost their crop and all their leaves, were not badly damaged. So it was short-term economic assistance that was required, together with building materials and labour to repair the damage to buildings.

I was fortunate in being able to benefit from three short-term sabbaticals during my time in the West Indies. The first was to the University of Cornell, New York State, to take part in its centennial celebrations. Cornell was the first Land Grant College to be established in the USA and has made a major contribution to agricultural development in the Third World. My visit coincided with the 'fall' and I shall never forget the brightly coloured maple trees as the leaves turned from green to red and purple almost before one's eyes. Soon after my arrival, I was asked by Professor Ken Turk, then Dean of the Faculty of Agriculture, to give a Guest Lecture to the undergraduates. I was surprised and slightly worried when about five hundred undergraduates turned up, including some stunning looking young American girls, who occupied the front row. My talk, I recall, was on 'Animal Breeding in the Tropics' and after the lecture I went up to the bevy of beauties who had been listening to my lecture. 'Tell me,' I asked, 'what is your interest in Tropical Animal Breeding?' 'Gee,' they replied, 'we know nothing about that. We just came to listen to your lovely English accent.' Ask a silly question, I thought, and you're bound to get a silly answer!

The second visit was to MacDonald College, Montreal. The Canadians were very keen on fostering close relationships with the newly Independent West Indian Islands and this was particularly true of the Province of Quebec, perhaps because of the historical French connection with the West Indies. They took the lead in setting up academic exchanges between the University of the West Indies and Montreal University. I spent several enjoyable weeks at both these institutions, giving a few lectures, conducting some research and having useful discussions with their animal scientists. While I was away, another aberrant hurricane was heading for Trinidad, causing much concern and anxiety. At the last minute its full force by-passed Trinidad and we only lost one small tree in our garden, but the hurricane caused great devastation in Tobago, especially to the coconut plantations Due to the depressed state of the coconut industry, the damage has never been made good and the scars of the hurricane still survive.

The final sabbatical was to Maracay Agricultural Station in Venezuela. The staff at Maracay had read about the projects I had carried out on the animal behaviour of tropical cattle and wished to repeat some of the work in their own country. I duly made several trips to the South American mainland to set up, and then to participate in, some of the experiments arising from this collaboration and the results were duly published (14). As with my students in Trinidad, the Venezuelan workers were fascinated by the finding that cattle only spend up to ten minutes a day actually sleeping. This fact, easily verified if one observes cattle as closely as we did, is not well known to 'experienced' cattle farmers. One of the ploys I used to attract the attention of a farming audience was to ask them how long they thought their cows slept each day. Most would guess several hours and few, if any, would get the answer anywhere near right. I was impressed, however, that several farmers took the trouble to check up on my assertion and have come back to me years later saying, 'You know what you told us about sleeping cattle. Well, although I didn't believe you at the time, you were right!'

In a small community like Trinidad one has to be prepared to turn one's hand to a variety of different jobs which would never come one's way in the UK. Two such jobs were offered to me, both with the media. The first was a daily early-morning 5-minute slot on Radio Trinidad called 'The Farming Programme'. Fortunately for me, I didn't have to be in the studio before breakfast every day, but about once a week I visited the recording studios and taped as many programmes as I had time to prepare, which were then broadcast each morning. The topics I covered were widely drawn and I got my ICTA colleagues to help me select relevant topics, which I then worked up into a 5-minute talk. When things went well, I had a queue of about a dozen programmes ready for broadcasting, but when things went badly, such as just after a holiday or an overseas trip, I was sometimes down to a single programme 'in the bag'. It was difficult to judge the 5-minutes allotted exactly and so I aimed to record 275 seconds of talking-time, and arranged that the technicians in the recording studio would play the credits and signature tune for a shorter or longer time, so that the final piece just fitted the allocated slot. I was paid the princely sum of $25 a programme so it was not exactly a profitable exercise, but it meant that I had to keep alert for suitable agricultural subjects to bring into the programme and I got a remarkable degree of feed-back from my farming friends.

The other job was with Trinidad Television, which was a new innovation at the time and very much in its infancy. I was asked to compère a weekly discussion panel called 'Time to Talk' . The time allowed was almost twice as long – 8 minutes – and it had to be broadcast live with no opportunities for rehearsals or repeats. I had to turn up at the TV studio 30 minutes before we were due to go on the air, together with a single interviewee or with a panel of up to four guests. There was a make-up artist who made sure our noses weren't shiny, but there was neither a Producer nor a Director nor any of the usual entourage which appear to be necessary to put such productions on today. The chief problem with these programmes was that my guest(s) would occasionally let me down by 'drying up' in front of the cameras, so that I was forced into giving a monologue for at least part of the programme. One could never tell when this would happen. Some of my 'prime' guest appearances would simply collapse under the bright lights of the television studio and the programme would fall flat on its face as a consequence. One of my best productions was when I interviewed four of my newly-arrived postgraduate students and asked them to give their first impressions of Trinidad. They were completely spontaneous and uninhibited and after the programme was over, the telephone lines to the studio were red hot with Trinidadian teenage girls asking for dates! I ran this TV programme for about a year before leaving the island and learnt a lot about getting people to talk in an interesting fashion with two large mobile cameras zooming on to their nervous faces. The pay for this job was slightly better – I received $200 a week for my amateur efforts!

My most challenging time at any University was the postgraduate period at Edinburgh, studying animal genetics. The reader may well wonder why none of my later experiments dealt with livestock breeding problems. The reason is that, before the days of biotechnology, when much of the research can be laboratory-based, livestock breeding was primarily a statistical exercise in mass selection, identifying superior males and mating them to large numbers of females. Such facilities were never at my disposal, either in Africa or in the West Indies, so I had to content myself with giving advice on animal breeding programmes. However, I did my best to keep abreast of such large-scale animal breeding schemes as existed at that time, and I wrote several discussion-type papers on my understanding of the subject in a tropical context. I also continued to take a keen interest after my return to the UK (28, 31, 85, 143, 151).

The social life in Trinidad was very much fuller than we had experienced in Uganda. There was a greater variety of things to do in one's spare time, and for much of the year we would visit one of the many nearby beaches and have impromptu barbecue parties in the evenings or at weekends. Local holidays were spent in the nearby island of Tobago, where we hired furnished cottages. In our latter years in Trinidad, we became very friendly with a marvellous Tobago farming family, Tony and Bunty Bishop and their five children, and we spent several very enjoyable holidays on their coconut estate near to Scarborough, the main town of the island. Tobago was a coral island, unlike Trinidad which was really a hunk of Latin America. The scenery was magnificent and, during our six years, the island was virtually untouched by tourism. The local population were friendly and once again one only had to buy the odd round of drinks in a rum shop and crack a few jokes to win their confidence.

It was very sad for me to see how much things had changed for the worse when I returned to Tobago in the mid-1970s. An enhanced political awareness and the effect of a booming tourist trade had made the local Tobagonians much more suspicious. The broad smiles did not spring so readily to their faces and I was shocked to learn that, during some of the 'political unrest', Tony Bishop slept with a loaded revolver under his pillow, fearful of attack by roving gangs of discontented folk intent on making trouble. As an agriculturist, it was even more depressing to see the standard of husbandry on most of the farms going steadily downhill. Coconut plantations were unkempt and untended, and the former high-quality cocoa and coffee farms had disappeared back into the bush. Unfortunately agriculture was no longer considered important and the status of farm workers became very low. Although lip service was paid by the Government to making Trinidad and Tobago self-sufficient in food, the actuality was that the population was becoming increasingly reliant on imported food. Unfortunately I see this pattern being repeated in Zimbabwe today under the misguided rule of Robert Mugabwe. Perhaps, with yet further difficulties put in the way of British farmers, a similar situation could even arise in the UK (see chapter 14).

One of my final lectures was delivered to a meeting of the Trinidad and Tobago Agricultural Society in Scarborough. The meeting was chaired by Tony Bishop, who was President of the Tobago branch of the Society at the time. I tried to convey the message that, in spite of lack of recognition by those in power, all people needed to be fed and Tobago was potentially one of the most fertile and productive parts of the world. Surely, I argued, this fact would eventually be recognised and agriculture would not only regain its previous prominent position in the economy, but would become even more vital to the well-being of the nation. Fine words, but unfortunately this new era has not yet dawned and Tobago, like so many other beautiful parts of the world, seems intent on exploiting the tourists and not inclined to respect and husband the land.

The longer I served in Trinidad, the more appreciative I was of the excellent work done by former ICTA staff, who conducted research under very difficult conditions in the early years of the College. I felt there was a real danger of such work being forgotten, especially as much of it tended to be published in minor journals which would not be widely read, I therefore tried to place some of these earlier studies on record. It is impossible to know how successful I have been, but at least I have made the attempt and done my best! (21, 31, 32, 33, 37, 51).

So our sojourn in Trinidad was interesting, busy and happy. Our family were growing up from babyhood through to primary and secondary school age, and when we left, the two eldest were going to boarding school in the UK. The students, as in Uganda, were as keen as mustard at pursuing their studies and were a delight to teach. The members of the mixed population were happy and contented and we greatly enjoyed joining in their respective festivals, such as the Hindu festival of lights – Diwali – and the big event of the Trinidadian year – Carnival – where most of the population joined in the two-day long festivities of *playing mas* (dressing up in elaborate costumes and dancing to the rhythm of the steel bands). The staff, with growing West Indian representation over the years, were happy and well motivated. The political changes, from colonial government to complete independence, took place without trouble of any sort and ICTA adapted to its new status as a Faculty of a newly independent West Indian University in an orderly manner. They were marvellous days and we look back at them with great satisfaction. I am very grateful that Sir Frank Engledow advised me, at our 1957 meeting on the verandah in Kericho, to 'Go West, Young Man'!

CHAPTER 9

Research at Colworth House

'How cam'st thou hither;
Tell me, and wherefore?'
William Shakespeare – Romeo and Juliet.

IN 1962 BUNNY'S FATHER died very suddenly. This made a big difference to our forward planning in many ways. Bunny's widowed mother found it increasingly difficult to cope on her own. Our eldest son, David, was not doing at all well in Caterham Junior boarding school and, having started near the top of the class, he was making his way fairly rapidly down to the bottom. Finally, although throughout our time in Trinidad there was a good rapport between the local and the expatriate staff, it was clear that, eventually, there would be mounting political pressure to 'West-Indianise' the University and I did not relish the idea of being around when moves were afoot to replace me. So, taking all in all, it was increasingly apparent to us that, with much reluctance, we would have to pack our bags and move back to the UK.

But returning home meant getting a job and there was quite a formidable 're-entry' problem for those with tropical overseas experience wishing to move back into British University posts. This was especially so at the higher levels, as thirteen years of tropical experience was not the best of backgrounds from which to secure a Chair at a British University. In view of these factors, it seemed fairly obvious that a move back home necessitated two major changes, the second being to a post outside the academic world.

On every leave I had spent time visiting University Departments and Research Institutions. Indeed, quite a lot of my 'holiday' period was taken up in this way. It so happened that, during a home leave in 1961, I had been interested to read in the agricultural press about the development of 'The Heavy Hog' by T. Wall and Sons, based on some interesting research conducted at the Unilever Research Laboratory at Colworth House, in Bedfordshire. Thinking this might be relevant to pig production in Trinidad, I duly wrote to T. Wall (and his Sons) and sought permission to learn at first-hand about their heavy hogs. I was duly invited to spend a day with them in London. I was right royally entertained (apparently my earlier work on growth and development had not escaped their attention) and both Bunny and I enjoyed the hospitality of their staff and went home laden with a large hamper of Walls' products. The final advice given to me was: 'If you wish to follow up this work, go to Colworth House, where you would learn about the research on which this project is based'.

It seemed that if Colworth could come up with another splendid day equivalent to the one in London, it would be an opportunity too good to miss. So I wrote to the contact I had been given, Mr Ken Dow of the Animal Science Division, Colworth House. If anything, the day at Colworth surpassed the hospitality in London and I returned home very impressed with the quality of the work being carried out, the ability of the scientific staff and the facilities at their disposal. Ken Dow's final comment to me as I took my leave was: 'If ever you think of leaving Trinidad, do write and let me know, as we might have a place here for you'.

So later in 1962, when I decided it was time to look for a post back home, I wrote to Ken Dow asking him whether this offer still held good. Back came a fulsome reply saying that the file had been passed on to the Head of the Cattle Science Department, Dr Bill Burt. Sure enough, a few days later Bill Burt wrote me a most kind letter, asking if I would be interested in accepting the job as Senior Cattle Scientist in his Department. It has never taken me long to make important decisions, so the following day I accepted the post, saying that I hoped to vacate my Chair within five months.

The last weeks simply flew by and the day of our departure arrived. There were numerous farewell parties, but perhaps the one which touched us most was a grand affair on the University farm given by the farm staff, numbering about 50 or so. One of the stockmen made a fine speech, saying that we would be long remembered and he produced the banner used at the trade union picket saying, 'More money for the Prof.' The staff presented us with several items, including a travelling bag, which was very well used over the next twenty years. Other farewells were of the more usual type and it was a source of much delight to us that all these events were truly multi-racial, without a trace of anti-expatriate feeling.

These events were not without their social difficulties. One party was given by the postgraduate students and two burly rugby front-row forwards from Fiji and Samoa decided to cook a suckling pig, Pacific Island fashion, as a special treat. This involved digging a large pit in our garden and filling the bottom with kindling and firewood. A layer of round stones was placed on top of the fire, which was then lit, so that the stones became white hot. The suckling pig, together with sweet potatoes, was wrapped in banana leaves and then placed on top of the hot stones and the whole excavation filled in. The

29. Winter Scene on the Colworth House Estate at Shambrook, near Bedford, 1965.

suckling pig was roasted for about twelve hours, when the pit was re-dug and the pig, now beautifully cooked, was solemnly conveyed on a wooden bier to the assembled party. Never has pork tasted so delicious and never has tender meat taken so much time and trouble to cook. The difficulty came in that some of those attending the party were of the Muslim faith and Muslims do not eat pork. The second and parallel difficulty was that the party was held in Lent, and the staunch Catholics present were in theory in the middle of a fast. Although I was unaware of the theological diplomacy going on behind the scenes, the party was a great success and much enjoyed by all present.

The final farewell event took place within hours – or even minutes – of our ship's departure. In order to pass through emigration, we had to produce a certificate showing that we did not owe any taxes, but the essential tax-clearance note somehow or other never turned up. In desperation I rang my friend, Mr Amoroso, the Government Commissioner of Revenue. He was a fellow member of the Port of Spain, Trinidad, Rotary Club. He told me not to worry as he would personally deliver my clearance certificate, but he wished to do this at his home, to which I would be driven by his official driver an hour before sailing time. Imagine our surprise when we found many of our friends, merry with rum, whisky and gin, holding yet another party in our honour. As always with Trinidadian hospitality, our host was very generous and the drink flowed ever more freely as the precious minutes ticked away. Gathering myself together, I begged for the vital Tax Certificate, which was eventually handed over within minutes of the expected time of departure of our ship. 'Don't worry,' I was told, 'The ship can't sail without a customs clearance certificate, which I have here in my pocket!' And so it was that we were oozed on to the ship, only to find another noisy party in full cry on board ship, waiting to give us a final send-off. The ship sailed slightly late that day and it has to be admitted that we did not take our leave of Trinidad with quite the same degree of formality as had been accorded to Lord Hailes a few years before.

A month later I reported for duty at the Cattle Section of Colworth House Research Station, to start a nineteen-year stint with Unilever. For my first year I was in charge of nutritional experiments with beef cattle under the supervision of Dr Bill Burt. The cattle section was sited on one of the outlying farms a mile or so distant from the main centre of operations at Colworth House. The isolation was apparent not only in the geographic sense, but also because scientists from the other sections at Colworth rarely visited Anthonie Farm, which was not renowned for being the most hospitable part of the Colworth establishment. I made friends with the staff of the other livestock units and did my best to keep abreast of their work as well as my own, since at the end of the day, the end-users of this research were eight Unilever operating companies who were interested in the collective output from all the animal sections. It seemed obvious to me that there should be many occasions where research conducted on one class of farm animal was of equal relevance to another. For instance, the Poultry Section was involved in experiments looking at the requirements for essential amino acids in the diets for chickens and although these results were not directly applicable to mature cattle, they were of some significance to young calves and piglets at the pre-weaning stage.

Because Colworth House had been set up on a 'commodity' (species) basis, these synergies were not always easy to identify. The operating companies, however, wanted the maximum possible benefit from the work at Colworth. Indeed, the senior technical personnel of the operating companies were responsible for the nutrition of all classes of livestock. They therefore appreciated it when the work was presented in such a way that all possible opportunities for exploitation were described. There were other barriers to the fullest possible use being made of the excellent work conducted at Colworth. In addition to the Cattle, Pig and Poultry Sections there were also large 'pure science' Departments of Animal Biochemistry and Physiology.

Again, the work was very departmentalised, and artificial constraints between disciplines were erected.

All this was a very big change for me, having been head of a Faculty of Agriculture which embraced all crop and animal activities. I was now a new boy in a relatively small Cattle Section, where working contact with the other Sections was limited. It was only to Bunny that I acknowledged I was finding the job a little frustrating. I did what I could to break down these boundaries and I received a great deal of encouragement from Ken Dow. After exactly one year, Ken called me into his office and said, 'Well, Peter, how are you getting on?' On my guard I replied, 'Fine, why do you ask?' 'Well', he said, 'I thought that you might be getting a little frustrated confined to just the Cattle Section. How would you like to head up a new Section with a brief to do the sort of work which is not being done in any one Section alone?' 'Excellent idea,' I said, 'When do I start?' 'What about tomorrow morning?' Ken replied and so the next day I found myself Head of the Biometrics Section, with a much more far-ranging brief. Bill Burt, understandably, was not too happy to see me go to this ill-defined new post. However, I set to work to show that my Section would bring benefits to the others. I was soon more than fully occupied with a variety of tasks, such as building and commissioning a carcass dissection laboratory, running a blood-typing laboratory for pigs, and taking charge of a punch-card clipping facility for our brand new Elliot computer. I recruited quite a number of new staff, which concerned me somewhat, as in Trinidad increasing staff numbers was no easy matter. I went to Ken Dow and asked him what he thought about my modest plans for staff expansion. 'Be reasonable,' he said. 'Fine', I replied, 'but what do you mean by reasonable?' 'Well', he said, 'two or three new staff every year is reasonable. Ten or twelve is not!' I calculated that, if I stayed twenty five years at Colworth and if I was 'reasonable', all the staff on the Station would be working in the Biometrics Section when I retired. That, I thought, was unlikely.

As time went by, the barriers between the various Animal Sections were perceptibly lowered. Perhaps this would have happened anyway, but I like to think that the example provided by the Biometrics Section demonstrated that there were tangible benefits to be gained from an inter-disciplinary approach. There were two other Sections similar to mine with wide-ranging briefs: the Analytical Chemistry Section and the Statistics Section. These had already operated successfully 'across the field', but in the past they had not been regarded by their 'pure' science colleagues as research workers. I tried to encourage these chemists and statisticians to innovate and experiment in their respective disciplines. As far as I could judge, they responded well to this, but whether or not they were still perceived as being 'second-class scientists' by their more purist colleagues, I never knew.

Although the job at Colworth only lasted for three years, we made many friends during that short period. When I joined Bill Burt's Cattle Section, one of the research scientists, Roger Dunton, took me under his wing and showed me the ropes. Our two families became very friendly. In the course of time the Duntons emigrated to Canada and are the proud owners of a small-holding in the northwest of that great country. Every year we receive warm invitations to visit them there and this we intend to do. Unfortunately most of my Canadian trips have been business ones; attending conferences and leading study-tours. When I started the new Biometrics Section, I recruited several staff from diverse disciplines. One colleague was a medical doctor, Joan Hardy, who was an expert in haematology. Part of my remit was to run a pig blood-typing consultancy service for one of the operating companies who were involved in pig breeding and Joan ran the laboratory, which we set up for this purpose. We still keep in touch, although our paths have taken us in very different directions since those far-off Colworth days.

Another close friend and colleague was David Hughes. I did not recruit David, as he was transferred from the Cattle to the Biometrics Section when this was first formed. David was an incredibly hard worker who thrived on difficult tasks and assuming new responsibilities. Another function of the Biometrics Section was the running of an abattoir so that beef cattle and pigs could be slaughtered under carefully controlled conditions and I refer to this work later in this chapter. We had to plan, build and then run this complicated facility and I put David in charge, after first sending him on a butchering course at the Smithfield College in London. The new abattoir was highly successful and a lot of time was spent showing meat scientists from other laboratories round it. David is still at Colworth, where I am sure his many talents are in as much demand as they were in my new Biometrics Section.

I have already mentioned that I made it my business to build bridges across the different sections of Colworth, which were somewhat distinct and introspective when I first joined. These led to the development of many friendships. In particular, two scientists in the Poultry Section became firm friends, Derek Stringer and Derek Shrimpton. Derek Stringer left Colworth to join ICI and was the scientist responsible for the work on their novel feed ingredient, *PRUTEEN*. He subsequently got a job with the European Commission in Brussels and is now very happily retired in the Lake District. Derek Shrimpton, like me, left Colworth to join a Unilever food company BOCM. He was the Chief Poultry Adviser to the company when I was moved to Liverpool as Agricultural Director of Silcock and Lever Feeds and when the two companies merged, in 1971, we were colleagues once more in the new BOCM Silcock headquarters. Derek Shrimpton was very much a 'Cambridge man' and has now retired to a country village a few miles west of Cambridge, where he is active in nutritional consultancy work.

I also became very friendly with the scientists working in the Pig Section of Colworth. The Section was headed by Bill Coey, a delightful Irishman who was a keen fly fisherman and an even keener trencherman. Bill was a charming and most co-operative colleague and was one of the most popular Section Managers on the station. Sadly, his 'problem' with 'the hard stuff' eventually became his undoing and Bill died in less than happy circumstances back in his beloved Ireland. I hope that there is some good salmon fishing, and some good Irish whiskey up in Heaven for Bill's benefit! Bill was very good at recruiting junior staff and two of his prodigies went on to do great things in the upper echelons of senior Unilever management. Two of Bill's colleagues from Colworth days joined BOCM Silcock, and were valued members of my nutritional group at Basingstoke, namely David Loane and Mervyn Eddie. Mervyn left me to become an Area General Manager and subsequently ran the Unilever Scottish Fish Farming operation, Ocean Harvest, and later took charge of Unilever's oil plantations in Malaysia. Mervyn and his wife Sandra have now retired to Edinburgh and we meet socially from time to time.

One of the many pleasures of my five years' stint at Colworth was being invited by the operating companies to act as a guest lecturer at agricultural meetings they had organised and at various other company events such as conferences. I enjoyed getting back to the grass-roots of the business, seeing how the companies operated and comparing their different ways of doing the same job. One had to be very discreet because all the companies, although in theory all members of one large Unilever family, were competing with each other in the field. It was a difficult line to draw when to say: 'Have you thought of doing it this way?', when the basis for the comment was a concept borrowed from a different company. A lot of diplomacy was involved, and one had to be very careful what one communicated to whom. Mistakes made in this context were difficult to rectify and confidences broken meant that one was effectively barred from receiving further sensitive information.

I also had to be very diplomatic in suggesting that some of the research conducted should be offered for publication in the learned journals. Most of our experiments were written up as *Colworth House Research Reports* which were rightly regarded as highly confidential documents. There was no way the operating companies wanted to see the results of work that they had commissioned given away to the opposition. On the other hand, it was important that Colworth was regarded as a responsible academic institution, both in order to recruit highly qualified staff and also to ensure that 'work from Colworth' cut ice when it was needed to support the patenting and licensing arrangements for new products. A fine line had to be drawn and some colleagues argued that if they never published anything, they could never be criticised. Happily my senior colleagues took the view that occasional publication was in the best long-term interests of Colworth and I did not have a problem in getting permission to publish several papers during my short period with Unilever Research (38, 40, 41, 42).

It was probably because I was able to adopt this more open approach that I managed to re-establish my contacts and friendships within the academic world. The first edition of my major textbook on *Tropical Agriculture*, written as co-author with Tom Webster, was published at this time (39) and this, together with other papers, brought me to the attention of Nottingham University. I was very flattered when, in 1967, they invited me to become a 'Special Industrial Lecturer' at the School of Agriculture. 'What will the duties involve?' I was asked by the Colworth Director. When I showed him my new textbook and said that I had been asked to give an annual series of ten lectures on tropical agriculture, he was reasonably satisfied.

I had not come across this problem of confidentiality whilst working in Universities overseas. There the whole ethos had been of a completely open academic environment, with ideas flowing freely in all directions. The commercial world brought the matter of Intellectual Property Rights (IPR) into sharp focus. In the late 1960s, the matter was not of much concern to most University Departments, unless they were working in very close liaison with private industry. The Agricultural Departments at that time were working on behalf of the farming community, or for government and so the realities of looking for patents and defending IPR were not then of much importance. All that was to alter. This early insight into 'scientific secrecy' was to prove of great help to me when I later returned to the academic fold and was able to help my colleagues understand the new environment in which they were operating.

One of the Unilever food companies was deeply involved in the poultry business. In order to secure their supplies of raw material, they invested in poultry-growing farms and set up their own chicken processing plant. The operation was successful, except that they had an embarrassing by-product of guts and offal from the factory, which became increasingly expensive to dispose of. I was invited to the company headquarters to see if I could suggest a possible solution to the problem. I tentatively suggested – with tongue in cheek – that consideration might be given to 'turning a stumbling block into a stepping stone' by feeding the poultry waste to mink.

Unilever operating companies are not slow to make decisions and next time I had contact with the company we found that they had set up a series of mink farms near to the poultry units and were successfully turning poultry offal into mink coats! However, a secondary problem had emerged. The mink pelts were fetching very low prices at the auctions, organised by the Hudson Bay Company, who had a monopoly position in respect of furs, including farmed mink. It transpired that there are 'fashions' in the colour of mink pelts, and that the producers of long-standing bred their mink a year or two in advance of market demand, knowing that their monopoly position could determine the 'correct' colour for any future season. This highly sensitive information was kept very much to themselves and any newcomer to the mink business was not privy to agreements concerning the future trends in pelt colour. The consequence was that Unilever was always out of step with market requirements, with the result that prices were not only low, but at or near the cost of production.

So both the poultry production farms and the mink farms had to go, and the company sourced its poultry meat from third-party growers. This episode proved a valuable insight, not only into the decision-making process of a large company, but also into the danger of seizing upon a new technology without a complete and thorough study of the possible consequences of changing the established method of doing things. With industry, the results of uncritical exploitation of new technology are always very readily apparent. In academic life the attitude is somewhat more relaxed.

A parallel case is also of interest, this time from the plant side of the business. Unilever is very heavily dependent on fats and oils, as they are the main inputs to its soap, detergent and margarine business. A traditional source of edible oil is the oil palm, grown in large quantities in various tropical countries, but most particularly in Malaysia. Unilever own large oil palm plantations and have a great deal of expertise in the growing of palm seedlings and in due time harvesting the resultant crop of nuts and extracting the palm oil. The problem has been that the trees in a plantation are all different. Traditionally, palms are grown from nuts which have been produced by normal sexual reproduction. If all the young seedlings were to be produced by vegetative reproduction, then in theory the plantation could produce a uniform crop of superior quality.

So Unilever Research spent time and effort perfecting a system of vegetative reproduction, carefully selecting good 'mother plants' from high quality trees. In a relatively short time many tens of thousands of plants were produced under controlled 'glass-house' conditions and in due course released into the field. In theory, they should have started to yield a year or so earlier than normal, but when the time came to expect the first crop of nuts, the palms were bare. In fact, the trees were sterile, for during the production process, a 'reproductive incompatibility' factor had been unwittingly introduced. The plantations had very fine stands of uniform, quick-growing trees, but unfortunately the fruits were nowhere to be found. A few small-scale experiments would have revealed the flaw in the operation and a lot of money would have been saved.

One of the major lines of work conducted in my new Biometrics Section was concerned with meat quality. These studies were carried out for the benefit of T. Wall & Sons, who were interested in the eating quality of fresh meat. Walls had a major abattoir in Atlas Road in North London, and a series of experiments were conducted in which the immediate pre- and post-slaughter regimes were modified, to enable the resultant joints to be more tender. The experimental treatments had to be carried out at the abattoir, but more detailed studies on the properties of the muscles needed to be conducted under controlled laboratory conditions at Colworth. This necessitated transporting a few cuts of meat, weighing about 20 kg, from Willesden to Colworth, a distance of about fifty miles. I had a small Transit van in my section, and was more than prepared to send the van to Willesden to fetch the meat, but the Transport Department at Colworth would not hear of it. This, they argued, was a job for the Unilever Transport Company, who had special refrigerated lorries for the purpose. A lorry was duly despatched and a couple of hours later I had a phone call in my office saying there was a very big vehicle trying to negotiate the narrow entrance to Colworth House. I drove to the main gateway to find a large articulated truck stuck at the gate. Inside was a small tray, on which my few modest joints were neatly placed. A few days later a large bill for the transport costs arrived on my desk, 'for the conveyance of 20kg of raw meat at exactly 1 degree Celsius'. Unilever never did things by halves!

I greatly enjoyed my four years in Unilever Research. Our three children were happily studying at very good schools in and around Bedford. We were successful in finding our first UK home. Bunny had become the Deputy Headmistress of a local girls' school and life was, as my brother used to put it, 'A bowl of cherries on a plate.' One day Ken Durham, the Director of Colworth, called me into his office and said, 'Peter, there is a job going in Liverpool. The Agricultural Development Director of Silcock and Lever Feeds is retiring. They have no acceptable internal candidates for the post and

have asked me to find a suitable candidate. I am putting forward your name.'
In 'Unilever speak' this question is equivalent to 'How soon can you sell your
house?' But first I had to gain the acceptance of the Directors of my future
company.

CHAPTER 10

A Director's Life

'A poor player struts and frets his hour upon the stage
and then is heard no more'
William Shakespeare – Romeo and Juliet

UNILEVER ACTS QUICKLY when it comes to filling posts at the Director level. A few days after being called into Ken Dow's office, I received a letter from Mr Francis Saint, the Managing Director of Silcock and Lever Feeds (SLF), a Liverpool-based company producing feed for animal consumption. The letter invited me to come to Liverpool for interview and gave me two dates from which to choose.

A few days later I presented myself in the magnificent head office of SLF, known as Stanley Hall. It was a 'custom built' stone edifice erected in the early years of the century. Francis Saint was a small, well dressed gentleman with piercing blue eyes which were fixed on you throughout the interview. He spoke with a clipped, public school accent and did not waste his words. Francis explained that, as he had just taken over responsibility for the new operation, there had been a 'significant change in management style'. Up to then, Silcocks had been run as a family business, the chairman being the third generation of Silcock owners, Dick Silcock. The old Silcock style was oak panelled rooms, light in appearance with traditional 'agricultural' pictures on the wall, such as good Constable reproductions. Francis Saint, therefore, had the light oak removed and a dark, almost black, panelling put up in its stead. Down came the Constables and up went impressionist paintings. Francis Saint had never been employed by Silcocks. He was a career Unilever manager, put in by the Unilever board to bring the Silcock family business firmly into the Unilever stable.

The interview went well. It was clear that my scientific credentials were taken as read (perhaps not surprisingly, as I had come from Unilever's central research laboratory). It was my managerial abilities that were under close examination. I liked the frank and forthright style of Francis Saint, who was clearly a man who knew where he was going and was keen to ensure that he had around him a team of Directors whom he could trust to get on with the job. In my case, this was the side dealing with agricultural science, advice to farmers and with direct responsibility for running the five company farms. Francis was very open about his own background. Soon after leaving school he was called up and served as a naval officer during the War. He had been torpedoed in the Mediterranean a couple of times whilst serving in destroyers. Since joining Unilever, he had made the usual progressive steps up the managerial ladder, one of his more recent postings being to South Africa, where he had been in charge of a Unilever-owned animal feed company.

Francis took me to the Directors' coffee room to meet his colleagues: Michael Abraham, the Marketing Director; Ray Harris, the Production Director in charge of the company's six mills; Harry Duerden the Finance Director and Dick Silcock, the titular company chairman, the grandson of the founder of R. Silcock & Son. The team members were clearly capable people, anxious to ascertain whether the 'new boy from Colworth' would fit in. I wondered what they thought of me and was intrigued to know how they would set about turning me down if I didn't come up to expectations.

After coffee I spent an hour with each Director, and then we all joined up again for lunch at the Liverpool Cotton Exchange building, sited immediately opposite Stanley Hall. The drinks flowed freely over lunch, but I noted with interest that, although everyone appeared to be refilling their glasses at frequent intervals, in fact they were only topping up with small quantities each time. The operation was clearly devised to see how well I could hold my liquor. In this regard my Trinidad days stood me in good stead. Drinks never flowed more copiously than in the West Indies, yet most professionals knew how much they had already drunk and how much more they could safely drink. In the afternoon a long series of meetings had been arranged with the staff who would form my Development Team, always assuming I got the job. I knew most of these folk already, as I had been at some pains to get on friendly terms with all the company agricultural scientists whilst working at Colworth. They seemed to want me to accept the post if it was offered. I suppose it was a case of 'better the devil you know...'

At the end of a long day, Francis spent another hour with me, spelling out his future plans and eventually made me an offer. I was disappointed. The salary envisaged was only about 15 per cent higher than that which I already enjoyed. I pointed out that a move to Liverpool would mean a change of house, a change of schools for the children and possibly the loss of a job for Bunny. All this couldn't be compensated by a 15 per cent rise. We negotiated for a long time, and eventually compromised with a pay rise of about 30 per cent, to be reviewed in six months' time. Some months afterwards Francis told me that, had I been minded to accept the first offer, it could well have

30. The author setting off to work in Liverpool as SLF Agricultural Director, 1968.

been withdrawn. Directors, he explained, would frequently be negotiating on behalf of their company. If they couldn't even negotiate on their own behalf, what chance did the company have to ensure advantageous business deals? I had not come across this ploy before. Academic salaries are pretty well fixed and it is usually a case of 'take it or leave it'. In business the rules are very different. If you aren't capable of negotiating, you are not seen to possess good managerial skills. Unfortunately, many folk learn this the hard way as no one tells you the rules of the game in advance.

For the next few months I commuted weekly between Liverpool and Bedford. The children all needed to complete their academic years and Bunny and I required time to find a new home. Motorways between Liverpool and the Midlands were still being constructed and the journey was long and tedious, especially when the trip started at about 6 p.m. and ended some four or five hours later. About every third weekend Bunny would come up to Liverpool and we would visit four or five houses which I had located as possible homes. It was very discouraging for me when the majority of the houses I had short-listed were written off on grounds of

31. The author building a patio at his Chester home, 1968.

external appearance. One very attractive house was rejected before she even saw it, as the smell of a nearby oil refinery hit us on our way to inspect it. I would have settled for about a dozen different homes before Bunny found the 'right' one, but her instinct was spot on. We ended up with a lovely house on the outskirts of Chester, which, in terms of construction and location, was probably the best of the 19 homes we have lived in during our life together.

Bunny never felt able to settle in Chester. Although she was happy with our new house, and the children were well settled in their schools, she seemed to know our stay would be brief. We could both see that, although Unilever had cut down the number of its animal feed companies from three to two, there was still a lot of unnecessary duplication between the two remaining companies, SLF and BOCM (British Oil and Cake Mills), and further amalgamation was inevitable. True, there was a difference of sorts. SLF sold directly via its own sales force of about one thousand company reps., whereas BOCM sold indirectly, through independent agricultural merchants and co-operatives. The 'party line' was that the farmers were either 'direct buyers' or 'indirect buyers' and we needed to supply both ends of a divided market. In practice, the farmers were looking for a high quality product at a reasonable price and it did not matter whether their feed was supplied directly by the company or indirectly through a third party.

I learnt a salutary lesson about the dependence of the company on the personalities of the haulage contractors, strangely enough from my hairdresser. Soon after taking up residence in Chester, I went for a haircut and when I sat down, the hairdresser said, 'You are Dr Peter Wilson of SLF'. 'How on earth did you know that?' I enquired. 'Your picture was in the *Farmers Weekly*,' he replied. 'What are you doing reading the *Farmers Weekly*?' I ventured. 'Simple,' he said, 'I keep pigs.' 'How many?' I asked. 'Two sows,' he replied. 'What do you feed them on?' I queried, thinking I might be in the process of making my first real sale, even if was only for two sows. 'Bibbys,' he told me. 'I used to buy SLF pig feed but the lorry driver swore at my wife and kicked my dog up the arse, so I told the SLF rep. I would never buy SLF pig feed again.' I was duly chastened but learnt an important marketing point: it is not the product alone which the customer buys, but the total sales package that goes with it. When you are a farmer, an essential part of the package is the courtesy of the lorry driver delivering your feed.

A few months later I came up against my first major problem. Part of my job was to give final approval for all changes in the specifications of the hundred or so products that the company manufactured in its six feed mills. There was, as always in such situations, a continuous dialogue between the company buyers on the one hand, who were hell bent on spotting 'good buys' in order to keep the cost of the products down, and the company nutritionists on the other, who were always trying to make better quality products. The buyers did not come under my line management but nevertheless I had the power of veto over their activities if I had good cause to believe that a batch of raw material would result in poor quality feed and hence lower performance.

The catch came in deciding whether a particular raw material on offer was 'new' (in which case it required my approval) or 'traditional', in which case it did not. One raw material which went into pig diets was haricot beans. Its quality was always checked by chemical and microbiological assay for pathogenic bacteria. The battery of tests was large and quite expensive to conduct, especially as some tests, such as those for the presence of poisonous weed seeds, could not be automated, and had to be conducted by hand.

In 1998, soon after my arrival, the buyers purchased a 50-ton lot of haricot beans at a knock-down price. The low price should have raised queries in their mind, but since the samples passed all the routine tests, no more questions were asked and the beans were duly delivered to a couple of mills and found their way into various brands of pig feeds. A day or two after the first deliveries to customer farms, our field staff received complaints from farmers that their pigs had gone down with acute diarrhoea. Instead of putting on weight, the pigs were losing weight at an alarming rate. Vets who were brought in to put matters right came up with one of their stock answers, 'Blame the feed', and in this instance they were right. Wherever the haricot beans went, the pigs sickened.

Now 50 tons of haricot beans goes into many hundreds of tons of pig feed, as its inclusion rate would be low. This meant that a very large number of pigs and pig farms would be affected. The marketing department panicked and gave instructions to the field staff to withdraw the offending batches of pig feed and settle for as little as possible – preferably offering the farmer replacement pig feed on a ton–for–ton basis. When I heard about it I was concerned to find out what had gone wrong, since the SLF pig feeds were made to a good specification; it was regarded as one of the best on the market at that time.

Even before the days of BSE, when 'traceability' became the order of the day, SLF had a good computerised detective system for putting together all the feed bag numbers (which were linked to the specific batches of ingredients) with all the farms experiencing pig problems and searching for the common factor. In less than 24 hours the answer was quite clear: pigs squittered when fed on feed containing the 'cheap' batch of haricot beans. There were still some unused beans available, so I ordered that all these be collected and sent to one of our research farms. This suspect feed was then made up into experimental diets at 5 per cent inclusion. Sure enough, within 12 hours of eating the suspect feed, the experimental pigs went down with severe diarrhoea. But how long would they take to recover after switching them on to a 'safe' pig feed? We experimented; some pigs being switched to normal rations after 24 hours feeding, others after 48 and yet others after 72. A few poor pigs were left on the haricot bean diets for as long as they would eat the feed.

From this crude experiment, we found out precisely the amount of 'damage' done to the pigs and therefore how much compensation was due for the time they had been fed on the haricot beans. I then attended an emergency board meeting and argued that, armed with this information, our reps. should visit all the affected farms and immediately offer a cheque for this amount in compensation for the damage caused. Not surprisingly, the Marketing Department was not too pleased about this, as it would clearly result in the pig brands showing a large loss on that month's account. However I was adamant that it was better to compensate before legal claims for damages started flowing in, rather than sit back and hope the problem would simply go away.

I won the battle and our reps. visited the affected pig farms. Within a week of the unfortunate introduction of the beans every affected farmer had received a prompt payment 'in full settlement'. Imagine the surprise of the Marketing Department when the accounts came in a month later. Far from a big loss, the extra business in pig feed sold later that month had more than wiped out the cost of compensation. Apparently the reaction of the farmers was, 'Here is a company that will see us right and offer us compensation even before we had got round to thinking about a claim. Not only that, but we saved a lot of veterinary and legal fees into the bargain'.

The unhappy sequel to this interesting little story was that we never identified the toxic ingredient in the batches of haricot beans, purchased by the company as 'a good buy'. I did find out, several weeks later, that the reason the beans were cheap was that they were what is known in the trade as a 'distressed sample'. In other words, they had been turned down by the manufacturers of tinned haricot beans for human consumption and had then been offered to the animal feed trade at a knock-down price. So in future any 'distressed samples' which the buyers wanted to purchase had to have my authority as a 'new' and not a 'traditional' raw material.

When I joined SLF, the Company owned and operated six fairly large farms. One was a training farm, where the company representatives (of which we had more than a thousand) learned their trade and the elements of applied animal nutrition. The other five farms were devoted to research and development trials. One was a pig farm in North Wales. Three were mixed farms, with pigs, poultry, dairy and beef cattle. The last was not so much a farm as a number of farm exhibits at the National Agricultural Centre at Stoneleigh, where we kept pigs, calves and beef cattle as a demonstration of a wide range of good farming practices.

I was responsible for what we called a 'Development Team', with well qualified advisers for all the three main animal groups, cattle & sheep, pigs and poultry, and with graduate development officers deployed on the farms, responsible for the conduct of all the experiments. At any one time some twenty or so development trials were in operation and each experiment was written up as a statistically analysed Development Report. I had in my team an excellent statistician, Paul Roberts; it was his job to ensure that all the experimental designs were sound before any trial commenced and to produce a proper statistical analysis when the responsible officer was writing up the results. I also had a first rate nutritionist, who by popular agreement was the best in the feed trade at that time, David Filmer (another former student of Wye College). Every month we held a Development Team meeting, with all senior staff present, when the Development Reports were discussed in detail. Francis Saint took a close interest in these reports. Although not a scientist himself, he always read each one and quite often the report came back to me with marginal comments and queries, some of them very apposite. I never had a single comment or query from any of my marketing colleagues, although their product managers took a close interest in our work. This was a great pity as we went out of our way to make the reports interesting and to describe any marketing angles that we thought arose from the work.

The output averaged about six fully documented reports for each team member each year. They were written up with the rigour which would have been needed for publication in a learned journal. Although used to contributing invited articles in the farming journals on matters relating to improvements in animal husbandry, my staff were not accustomed to writing up papers in a form that would satisfy the critical eye of a science journal editor. I therefore led by example, publishing five papers during my three years based in Liverpool. (42, 43, 44, 45, 46). However, the results of my efforts were rewarded later, when many more papers were published by my colleagues during my thirteen years with BOCM Silcock.

At SLF the constraints were less irksome than they had been at Colworth. As the Agricultural Director, I took full responsibility for authorising the publication of papers. This procedure resulted in the Development Team members becoming respected colleagues in the agricultural scientific community. They were encouraged to go regularly to technical and trade meetings and the fact that they presented papers meant that they tended to be envied by their colleagues in the other animal feed companies. Most of these rival colleagues were effectively prevented by their companies from talking about their work and consequently the scientists employed in Universities and Research Institutes did not have such a good rapport with them.

The Development Team was widely drawn, but it was made clear that contributions at team meetings had to be kept short and to the point and

32. The author addressing a Scientific Conference, Rotterdam, 1969.

that any major topic for debate should have been prepared well in advance. David Filmer was a key member of the team. He was exceptional in that he was equally well versed in all the animal species with which we were concerned. David was by far and away the best applied animal nutritionist in the country at that time but, like many outstanding specialists, he had difficulty in communicating his ideas to a non-scientific, non-farming audience. This included the Directors and the Heads of the other major Departments of the company. He knew farming inside out, and this was recognised by the field staff. They found that David was a skilful platform speaker, who could get his ideas over to a farming audience without difficulty. The senior management of the company were less appreciative, since they regarded David as a constraint on their own activities, which could have been so much more easily pursued, had it not been for the need for ever higher levels of nutritional exactitude.

When I came to SLF, David was really the only trained nutritionist on the staff. The other members of the team were contributing more in the way of husbandry knowledge than in the chemistry of advanced animal nutrition *per se*. Their views were nevertheless very important, but there was a need to provide a scientific challenge to David Filmer and we both agreed that a high priority was the recruitment of at least two more nutritionists, the new members of the team specialising in a single species. It was not until very late in the day, however, that I was able to put this plan into full operation. Eventually we managed to entice Colin Fisher and Peter Cowdy to join the team, which then became a winning combination of which the rest of the trade were rightly envious. The other members of the team were the two principal farm managers: Bill Thickett in charge of the three SLF experiment farms in Cheshire, and Geoff Crookes, the manager of the Inglewood demonstration and training farm on the Wirral. In addition, we had a group of Experimental Officers, of whom the most notable were Peter Kenyon (a pig specialist) and John Pearson (eventually responsible for our poultry units). The final member of the group was an outstanding statistician, Paul Roberts. Paul, however, was not just a 'number punching statistician' but a real innovator, responsible for many of the computerised management systems which we introduced to our customers as a means of enhancing their profitability.

In 1970, after I had been in the Director's hot seat for two years, it became clear that the animal feed trade was going through a very tough period. I was able to observe this not only as a member of the SLF Board, but also because I could gauge the overall state of the UK animal feed industry through my membership of the United Kingdom Agricultural Supply Trades Association (UKASTA), which was the recognised trade association of the animal feed business. Other companies were going through a more difficult patch than SLF, as inroads were being made into their market share by the growth of the agricultural co-operatives. A further factor was the increased amount of 'home mixing' by farmers making up their own rations.

It was also apparent that the strategy by which Unilever operated two separate feed companies, one selling direct to farmers whilst the other sold indirectly through agricultural merchants, was no longer sustainable. There was a great deal of duplication of effort, particularly in areas such as mine. The 'on costs' of our research and development programmes, and of our pool of nutritional expertise, added up to approximately £1 per ton of feed produced, and this at a time when profit margins were on average less than £1 per ton.

Thus it was no great surprise when a merger between SLF and BOCM was suggested. In their heyday, the two companies produced just under three million tons of animal feed, 1 million in SLF and 2 million in BOCM. However, Unilever were not very interested in volume; the parameters by which we were judged by our masters on the Unilever board were profit and return on capital. Both of these critical indicators were in steady decline and there seemed no way of reversing this long-term trend. To complicate the picture further, many of the 'small country compounders' were in the hands of private families, who were much less interested than Unilever in profit and return on capital. As long as the principals of these small firms earned a good living, with houses, cars and hospitality subsidised by the firm, they were quite content. Most of them came from farming stock and in many cases the owners of these small country mills were also farming in their own right. With no threat to their outward appearance of wealth, the normal economic pressures borne by a public company with shareholders to consider did not concern them.

Thus it was with a sad heart that I left my home at 5 a.m. on a warm summer's day in 1970 to visit three of the five SLF farms, with the task of telling the staff that the farms were being closed and their jobs were being made redundant. We had prepared for this moment for six months. The total number of farm staff affected was about one hundred. I held an open meeting on the first farm at 6 a.m. and then interviewed all the staff in turn, in company with their farm manager, explaining their personal position. Their pension, severance and redundancy money had all been worked out beforehand. I had personal letters for each member of staff, signed by the Chairman, to hand over at the interview. Wherever possible, the staff were offered alternative employment within the company, but as this necessitated a move, few of them were really interested. I had authority from the board to make extra *ex gratia* payments where special circumstances were brought to my notice at these interviews.

I arrived home that evening at 11 p.m. completely physically and mentally drained, but the job had to be done and it was my duty to carry it out efficiently and with humanity. The response of the men was remarkable and unexpected. On two farms, after my general talk to all staff setting out the position, I was publicly thanked by someone about to be made redundant for the full and frank explanation, which they understood and respected. On no farms were there angry outbursts or harsh words spoken. In 1970 redundancy was a relatively new phenomenon and none of the men I spoke to had experienced anything like this before. Yet to a man they accepted the situation with sadness but good grace and I felt very humble at the dignity and courage displayed by those who, to some of my senior colleagues back in Liverpool, were 'just simple farm folk'. I little realised that 'down-sizing' and telling staff they were to be made redundant was to be an unfortunate part of my duties for the rest of my career.

On this occasion, I only had one unhappy incident where the redundancy situation backfired. This was some six months later, just before the period of notice expired. The company had laid on its usual Christmas parties for staff deployed in the different regions. Not only had my farm staff been affected by the redundancy situation, but mill and sales staff were also involved. Each of these area-based events was attended by one of the Directors. At one party, where I was representing the Board, I was accosted late in the evening on the dance floor by a very drunk lady who turned out to be the wife of one of the redundant salesmen. She demanded to know why the board had picked on her husband. 'Didn't I know,' she screamed, 'that her husband was the best in the business? What had he done to deserve such terrible treatment?' I was ill-prepared for this onslaught, which compared so markedly with the civilised way in which my farm staff had faced up to this situation. I noticed, however, that she was not being supported by her husband, who kept in the background and looked as though he wished the floor would give way and swallow him up. I learnt from this that whereas the men involved (and in those days the SLF staff were 95% male) could understand the reasons for the crisis and could come to terms with it, the wives found the situation much more difficult to accept. Should we, perhaps, have arranged meetings with the wives of the affected staff? It probably would not have been much good, but I agonised for a long time over what more the company could have done to ease the pain of redundancy for the men's families.

The following spring SLF ceased trading as an independent operating company and merged with its sister company BOCM to form BOCM Silcock. Those of us who kept our employment, if not our old jobs, were transferred to the new headquarters at Basingstoke. The siting of this HQ had caused much debate and outside consultants had been called in to suggest the most suitable location. They came up with two names: Leamington Spa and Milton Keynes. A wise cynic noted that seeking the advice of an outside consultant was a waste of time and money. 'What they should have done,' he said, 'was to split the distance between the homes of the Chairman and the Vice Chairman and locate the new HQ there.' The then Chairman lived near Farnham in Surrey and the then Vice Chairman near Reading in Berkshire. Basingstoke is situated almost exactly between these two towns!

CHAPTER 11

Mergers and Take-Overs

'O let us be married!
Too long we have tarried'
Edward Lear – Nonsense Songs

IN 1971 WE HAD TO pack our bags and move house yet again. I was lucky to retain a job in the newly formed company, BOCM Silcock, as everything had been 'down-sized' and there was no longer a seat on the Board as Agricultural Development Director. In fact, there were no qualified agricultural directors at all in the new organisation. Francis Saint, the chairman who had appointed me and whom I greatly respected, was given the post of Marketing Director and the top job was taken by my former boss at Colworth House, Ken Durham.

The Chairmanship of BOCM Silcock was Ken's first company posting. Up to then his career had been in Unilever Research and I have already recounted how Ken had personally put my name forward as a potential Development Director for SLF. Ken was one of the cleverest and most intelligent persons I have ever met, and certainly would stand out a mile in any organisation. After leading BOCM Silcock very successfully for five years, he went on to become a main Board member of Unilever, eventually becoming Chairman. Knowing that he lacked the usual company background, he went out of his way to balance his team accordingly, with people strong on the commercial, marketing and production sides. With his own expertise in research, he was quite confident at looking after the scientific aspects of the business and so he did not see the need to have a qualified scientist as Agricultural Director.

Although there was no board job for me, I retained my 'pay and rations' and became first the Chief Cattle Adviser and then, a few years later, Chief Agricultural Adviser of the new company. Although naturally disappointed, I was content to make the best of the new situation and was encouraged when Ken took me out to lunch and explained that the company needed someone to 'portray an external image of scientific excellence'. I was given *carte blanche* to get out and about, representing BOCM Silcock and making sure that we had a good name in both the farming and the academic world.

The new portfolio was quite a challenge, but I had a most excellent colleague to work with in the person of John Deeley, our very capable Public Relations Officer. John had held this same job in SLF and brought the style and title with him from Liverpool. He and I together had a fairly robust budget and we made an unique team in the agricultural world. Between us, we initiated several high-profile activities, which sought to bring BOCM Silcock closer to the farming community. John organised an extensive series of working dinners with the leaders of the business communities, many of whom he knew well from his days in SLF. I took responsibility for similar high-level meetings with members of the scientific community with interests in agricultural research.

At one of the first of these 'boffins' dinners' (as John Deeley described them), the Secretary (Chief Executive) of the Agricultural Research Council was our principal guest. The post was held at that time by Sir William Henderson, a distinguished vet who had previously been Director of the Compton Research Institute. Gregor Henderson (as he was known) had not been so well briefed by me for this occasion as he should have been. Before the dinner party, he made enquiries through his PA as to what the 'hidden agenda' was for the meeting. When told it was simply an informal 'getting to know you' occasion, he was very perplexed, since senior civil servants are brought up on the theory that 'There is no such thing as a free lunch'. At the end of the meal I could not resist teasing him and making the point that the dinner had indeed been completely informal and that we had not used the occasion to lobby him on any particular issue. He replied saying that he now appreciated this, but it was the first time he had been wined and dined so well without some ulterior motive.

During the early years after BOCM Silcock had been formed, John Deeley and I planned the formation of two complementary, but different, Clubs. One, which John organised, was the 'Three Hundred Cow Club', which was an informal gathering of farmers with large herds of dairy cattle, namely 300 or more. The Club was serviced by the Company but run by its members, who numbered about one hundred. They met twice a year for a farm visit, followed by a convivial dinner-discussion in a convenient hostelry. This Club still exists, although its connections with its original host company have now been severed. The companion Club, which I set up, was the 'Three Thousand Gallon Club', which drew its membership from scientists, economists, veterinary surgeons and others, each coming from a different discipline. This Club also met twice a year, when visits were made to Research Institutes or University Departments of Agriculture, also followed by a good meal and an after-dinner discussion on a topic connected with the question, 'What are the factors responsible for the expression of high yields in dairy cows?'. Again, this Club is still functioning, although in recent years there has been a reaction against the old name so it now operates as the 'Dairy Science Forum'. Membership of this Club has been limited to a maximum of 25, and some of the after-dinner talks and discussions have been of exceptionally high standard. The meetings are unique in that, as no two members come from the same background, a common language must always be used so that every member can contribute fully to the debate.

After the Club had been meeting for about a decade, the members thought that there had been a surfeit of 'star-gazing' and not enough down-to-earth activity. It was decided to undertake a survey of the highest-yielding dairy cows in the UK, to see if there were any common factors. Finding the high-yielding cows was remarkably easy. The Milk Marketing Board of England and Wales kept computer records of all cows which were 'milk recorded' for purposes of breeding and management. As one of our members, Dr Peter Wood, was the Chief Statistician in the Milk Marketing Board, he obtained a print-out of the cows which, according to their records, had produced 750kg of 'milk fat plus milk protein' in a single lactation.

The plan was to visit all the farms which had two or more cows which came up to this exacting standard. Firstly we had to get the permission of the farmers to use their records and to make the visits. This was easily achieved. Out of 30 or so farms, only two owners objected or wanted to impose conditions which would have put the survey in jeopardy. The devising of the questionnaire to be used during the course of the visit proved a much greater problem. Three Club meetings, spread over 18 months, were needed to agree on the wording of the questionnaire, after which the visits were undertaken in pairs, such that no two members of the pair were in closely related disciplines. I was paired with Peter Wood and together we visited three farms.

The results of the survey were published as two separate papers, one dealing with the statistical findings and one with the more subjective management results (135, 136). The results surprised the research scientists within the Club, but not those members whose jobs brought them closer to the realities of dairy farming. There was only one common factor which applied to all the high-yielding cattle, namely that they were all present in herds where management was of the highest standard. In these herds the cows came first; the rewards and life styles of the milkers and herd managers a very poor second.

It followed from this main finding that breeding, feeding and management went together and it did not seem to make any difference whether the cows were fed on hay or silage, whether they were fed twice or thrice a day, whether the grazing system practised was paddock grazing or strip grazing. As long as the management of the chosen system of dairy farming was first-class, everything else followed. The answer to the subsidiary question was also simple. The top cows were found in the best herds, although the range of milk yields in these herds was wide and the average was well below that of their very best animals. There were no high-yielding cows in low- or average-yielding herds.

It might be thought that this finding did not tell dairy farmers anything which they didn't know already. Nevertheless they were keen to be told the results of the survey and three meetings, open to all the co-operating farmers, were held in different parts of the country. These meetings provoked the most lively and informative discussion. Some of the farmers involved in the survey were also members of the Three Hundred Cow Club and we duly received an invitation from our sister Club to present the results of our work at one of their meetings. Again, the interest was very great and a wide-ranging discussion followed the formal presentations.

Because of the success of these two clubs, I tried to form two parallel clubs, one dealing with pigs and the other with sheep. The pig club was given the name of the 'John Hammond Pig Group', since readers will recall that John Hammond was one of the father figures in the animal science world and much of his scientific work had been conducted on pigs. This club was successful for about twelve years, after which it appeared to run out of steam and so that eventually the decision was taken to wind it up. The sheep club never got off the ground. Although there was strong support from a few sheep scientists, there were never enough enthusiasts to form a critical mass and so the idea of this fourth club was tactfully dropped.

The two other species of livestock, poultry and turkeys, already had their own successful groupings. Thus the poultry farmers organised a National Poultry Industry Conference each year and I was invited to be its President for two successive years, when the Conference was held at Blackpool. I also addressed the Conference on several other occasions (149, 153, 164). The poultry people were different in every respect from dairy and pig folk. Poultry farming had by that time become 'big business', where attention to financing, marketing, housing and environment control were more important than attention to the needs of each individual hen. The result was that the industry was dominated not by husbandry-men but by agri-businessmen. They were commercial managers, and when they met they talked the language of the business world. Gross and net margins; return on capital; constraints to marketing; the identification of new product lines and novel market niches. When papers on management and feeding were built into the Conference programme, attendance dropped away as the delegates did business in the bar. Here the talk was of mergers and take-overs, rather than of protein requirements or disease control. These more scientific subjects were dealt with by the 'poultry boffins' at their own series of meetings, organised by the World Poultry Science Association. I attended these meetings, as well as the Blackpool Poultry Conferences, and you only had to eavesdrop for 30 seconds to know exactly which meeting you were at. How different from the days at Shadrack Farm, where I searched for clutches of eggs laid in the hedgerows, or at Fawkham Green, where Ken Smith taught me how to kill and dress a capon or truss a turkey.

The turkey folk had their own separate organisation, the British Turkey Federation. They also organised annual conferences as well as regional annual dinners, where the delegates came with their wives to listen to a paper on a turkey subject and then enjoy a four-course turkey dinner. One year I was invited to be guest speaker at the Norfolk Turkey Dinner. I was glad that the talk came before the food or the audience would have been very soporific. Choosing a title and style that would appeal to the ladies was very difficult, especially when it is realised that these turkey farmers were not really farmers at all, but businessmen in disguise. Many of their wives had probably only seen a turkey on a dinner plate, and few of them would have set foot inside a turkey house. I chose the subject of 'Turkey through the ages' and described how wild turkeys had been an essential part of the American 'Thanksgiving' celebrations long before they replaced the goose as the main item on British Christmas menus. I kept it short, since 'It is a foolish man who stands between a herd of wild buffaloes and their waterhole!'

After my speech, the turkeys were trundled in, one large turkey to every table. Each table was presided over by a member of the Norfolk Turkey Farmers committee, dressed appropriately in a chef's hat and apron. I watched with amazement the ensuing carnage of the turkeys. Out of twenty carvers, only two or three knew anything at all about the art of wielding the knife. Most attacked the turkey with such a savage onslaught that one wondered if they thought it was still alive and needed killing. Others ended up with a pile of turkey chunks which looked as though they had been flayed rather than carved. It was a most dramatic demonstration that the men wearing the aprons had probably never learnt to carve turkeys in their life and certainly had never been shown how to pluck and dress them. I was glad that Ken Smith was not there to witness the spectacle, as I am sure he would have been horrified.

In addition to leading the BOCMS 'Development Team' and taking a close interest in all the research conducted on our development farms, I tried to interest myself in some more general problems which required thought and discussion but were not themselves subject to experimental investigation. The first area of interest was Biological Efficiency. By this I mean the ratio of input to output of various biological entities, such as energy and protein. The subject is important, since if ever some of the major inputs becomes limiting, then those animals and systems which have the best conversion of feed input to edible output will tend to be favoured, whilst those with low efficiencies will be at a disadvantage. My contributions in this important debate have been made in various ways, from lectures to the Royal Society of London (49, 142, 167) to chapters in books (47, 59, 113, 156) and individual papers in a range of journals and conference proceedings (42, 45, 55, 65, 144, 155). Related to these studies is the consideration of the inherent competition between human food and animal feed. It is clearly far more efficient for man to consume vegetable crops directly, such as bread from wheat, rather than to feed crop products to farm animals (such as giving barley to fattening pigs). On the other hand, there are many crops which cannot be directly consumed by man, such as grass or cereal straws, although certain classes of farm animal can make good use of them (67, 75).

The last area of interest was relatively new when I first involved myself in it, but over time it has became increasingly important. This is the field of animal welfare. Thirty or so years ago practices such as the production of veal in crates, the penning of pigs in sow stalls and the keeping of five or more hens in a single battery cage were commonplace and rarely challenged. Now, however, such agricultural methodologies are either unacceptable to the discriminating consumer or are banned by law. In the 1970s few farmers took these considerations very seriously or, if they did, the constraints were regarded as gross interference by a crankish minority of the population. I regarded it as highly probable that farming would have to adapt so as to accept new restrictions on the way livestock were managed and I thought it necessary to try to convince farmers that change was in the air (92, 123, 129, 145, 147).

I was fortunate in that I continued to lead a small but excellent team of nutritionists and advisory specialists during my years at Basingstoke with BOCM Silcock. I was unfortunate in that during my thirteen years there I served under five different Company Chairmen. Partly as a result of this, the structure of the marketing side of the company was constantly being reorganised and redeployed, with consequent disruption to the inter-departmental working relationships. It seemed that each Chairman had to make his mark on the Company by a major re-grouping of the staff, but no sooner had the new structure got going than the next Chairman came into the picture, resulting in yet another round of 'musical chairs'. As with the party game, each new arrangement meant that some good colleagues were surplus to requirements and had to take early retirement.

One of the operational areas of the company, Scotland, seemed immune to these processes, although even north of the border there was a certain amount of 'down-sizing'. Jim Armitage was the Scottish Area General Manager for the whole of this period and was unique in this respect. He fought his corner very hard and kept most of the Scottish structure intact, most of the time. Jim was very well respected, not only in Scotland but in the Company as a whole. This enabled him to claim that, although changes might be necessary south of the border, they were not needed in the far north. He played the Scottish card very well each January, when he organised a magnificent Burns Supper. I was a regular guest on these splendid occasions and Jim always invited the Chairman and members of the Board as his official guests. The evening was a long one with the list of toasts rivalling in length the 'Bill o' Fare'. All the speeches were delivered by members of the Scottish Area team and were of exceptionally high quality. The timing was immaculate, with each speaker given a predetermined period for his contribution. Woe betide anyone who trespassed more than 30 seconds into injury time! I think the Directors from Basingstoke were so impressed with this display of well-orchestrated team spirit that they decided on their return to leave the Scottish Area structure intact for yet another year.

Scotland, unlike all the other Areas of the Company, had its own Agricultural Adviser, Alan Harker, who reported directly to Jim Armitage. This could have been difficult since the relationship of the Scottish Adviser with the Company Chief Adviser was not clearly spelt out. Thankfully Alan was an exceedingly able and sensitive person and he and I got on well together without our formal relationships ever being officially laid down. I was a frequent visitor to Scotland, at Alan's invitation, and greatly enjoyed the many trips I made to farmers' conferences and staff meetings in Bonnie Scotland.

My diary of engagements during my time in Basingstoke was very full. As a result, I was only in my office for at most a couple of days each week. This meant that I relied very greatly on staff looking after my work during my absence and keeping me informed if an unforeseen crisis cropped up. I was exceptionally well served by my close colleagues and especially by my Technical Assistants. The first of these was an outstandingly able agricultural graduate, Gay Climm, who subsequently married a contemporary at University who farmed in the Welsh borders. The second, who served me for over ten years, was Tim Brigstocke. I had several secretaries, the best of whom were quickly spotted by senior colleagues and unashamedly 'pinched' from under my nose. In my last years I found a winning combination; a Secretary who would not work for the company in full-time employment but only as a 'temp'. The reason was personal; Margaret Ireland was married to a busy lawyer and whenever a gap occurred in his business schedule, he took a holiday in France, and Margaret went with him. As a full-time member of staff, Margaret would not have been allowed this flexibility. Staff holidays had to be pre-arranged several months in advance. I was naturally somewhat inconvenienced when these French holidays cropped up, but I was secure in the knowledge that for the rest of the year I had the best secretary-cum-personal assistant in the whole of Basingstoke. Indeed, I knew this to be true, for many of the members of the Basingstoke Rotary Club had had experience of Margaret Ireland's exceptional clerical qualities and had tried, unsuccessfully, to engage her full-time.

One of the innovations introduced by my PR colleague, John Deeley, was a National Competition to find the 'Dairy Farmer of the Year', later modified to the 'Dairy Herdsman of the Year'. A team of national figures, highly respected in the dairy industry, judged the competition and were flown by helicopter, at Company expense, to make on-site inspections of the farms/herdsmen under consideration. The BOCM Silcock staff took no part in the judging, which, for many years, was conducted under the watchful eye of Sir Emrys Jones, Chief Scientific Adviser to the Minister of Agriculture. John Deeley and I came into the picture when the prizes were awarded at a magnificent luncheon at the Cafe Royal Hotel, London, where all the 'good and the great' in the dairy farming world were invited as guests of the company.

John Deeley insisted that a 'well-known public figure' should be invited to present the awards, which took the form of a splendid cup and a large cheque. I was less enthusiastic, thinking that 'public figures' would cut little ice with members of the farming community. I was wrong. On one occasion, the person John selected was Anna Ford, the television news announcer, and I hosted the table where the guest was seated surrounded by the three prize winners and their wives. Anna Ford proved to be a real professional. She had no difficulty in establishing a close rapport, not only with the award winners but, perhaps more importantly, with their wives as well. She revealed that part of her childhood had been spent on a farm and that she had actually hand-milked a cow. What I hadn't realised was that this was the first time many of these ladies had ever been to London, let alone been entertained by a television personality. It was clearly an experience that

33. The author with Anna Ford, Chief Guest at BOCMS Dairy Herdsman of the Year Awards Ceremony, Cafe Royal, London, 1981.

they would never forget. John had been absolutely correct in his choice of speakers and the ceremonies were all highly successful. It was my job to act as MC on these occasions and to say some appropriate things about both the award winners and the guest speaker and it was not easy to follow a true professional like Anna Ford.

As the Company's Chief Adviser, I was involved in a large amount of lecturing and company-sponsored farm demonstrations up and down the country. During the winter season I was on some platform two or three times a week. These events were organised by the Company field staff and I was simply told the venue of the event and the chosen subject. Now it is very difficult to be completely up to date with developments in all the different branches of the livestock industry. Things change, and change fast. However, I was fortunate in having a first rate team at Head Office, who were specialists not only in nutrition, breeding, farm costings, economics and farm management. I also had the managers of the four Company farms to keep my feet firmly on the ground. Perhaps most important of all to me at that time was my Technical Assistant, Tim Brigstocke. I had first come across Tim when he was President of the Seale Hayne Students' Agricultural Club and my host at a lecture I was invited to deliver at the College. He was clearly an outstanding student and at that stage he had no definite career plans.

A short while after giving my talk at Seale Hayne, I had a vacancy for a Technical Assistant and immediately thought of Tim, who accepted the position by return. Thus began a very productive partnership which lasted about ten years, during which period Tim became responsible for helping me with my extensive lecturing programme by researching the subjects, gathering statistics and drafting my talks. After a short while he was so good at the job that I included his name on any papers we subsequently published. In the course of 10 years in harness, Tim was named co-author with me on no less than 34 papers. This included helping me prepare the second edition of my textbook on *Tropical Agriculture* (110), where a great deal of work was required tracking down the libraries which contained the papers I needed. A little later we jointly wrote a text book on *The Improved Feeding of Cattle and Sheep* (126). In spite of his sterling qualities, I found it very difficult to persuade the company to promote Tim as he deserved, but he loyally stayed on as my right-hand man. Hence I was absolutely delighted when, a short time after I left BOCM Silcock, Tim was at last given due recognition and promoted into the position I had previously held. Assistants such as Tim are invaluable in supporting folk such as me, who have undertaken, or been given, far more work than can be undertaken single-handedly.

However, I was not entirely without knowledge of the subjects I spoke about! My Edinburgh training in genetics meant that I had no difficulty in holding my own when animal breeding was on the agenda. My first postgraduate work had been with poultry and overseas I had worked with dairy and beef cattle and with the East African dwarf goat. My own research had been in the field of basic animal nutrition and physiology and so my experience before taking the post of Chief Agricultural Adviser was fairly wide-ranging. It may be thought that African goats were somewhat remote from British sheep, but fundamentally there is not a great deal of difference between the two species. However, on one occasion at a sheep meeting in the far north of Scotland, I found myself a little out of my depth. I had a complicated itinerary and my plane was late. Partly because of this, I arrived at the meeting very tired with no time to spare. The lecture did not go down as well as I had hoped, and at dinner afterwards I apologised to the farmer who had chaired the meeting, a very prominent Scotch Blackface breeder.

'Dinna blame yersel', laddie', he said, 'Blame the silly b——r who invited ye!'.

Some of the most enjoyable of these farmers' meetings were those held in the Channel Islands. These islands are very small and the farming community few in numbers. More to the point, meetings with an 'outside' speaker are few and far between, with the result that almost the whole of the farming community turns out on these occasions. These dairy farming audiences are some of the most appreciative gatherings I have ever addressed and I tried to get to either Jersey or Guernsey at least once every year (70, 101, 114, 124, 138). There were a few tricks which one had to have up one's sleeve. In the Channel Islands, milk yields are measured not in pints, pounds or kilograms but in 'pots', and land is not divided into acres or hectares but into 'vergees'. These are traditional French units, and since they are in common use, I always had a 'crib' with me so that when lecturing, I used the appropriate local units and not 'foreign terms from over the water'.

The Company also had branches of the business in both Northern Ireland and Eire, and I had to visit our field staff and address farmers' meetings on both sides of the border. These were interesting meetings, also very well attended, where an appreciation of some of the finer points of Irish humour was desirable. In Northern Ireland it is quite acceptable to tell jokes about the Southern Irish. In the Republic of Ireland it is all right to poke fun at the 'Kerrymen' in the south-west of the country. In County Kerry, if one is rash enough to speak there, the only safe policy must be to tell jokes against the English.

On one occasion I was asked to address a large audience at Jury's Hotel in Dublin. I was told the meeting would start at 8 and that the lecture would be preceded by a dinner with some leading lights from the Irish farming world. I arrived at 6 and was told that I would find my dinner party at the bar. They were already assembled in large numbers, and were already well into their second or third round of drinks. At 7 we were still drinking. At 8.30 I was getting hungry but, more to the point, I was worried about the audience, which by then should be waiting for me. I slipped out and looked into the large room where the meeting was supposed to be taking place to find three Irish nuns knitting in the front row. A little non-plussed, I asked them if they had come to the lecture and they replied, 'Yes, and you must be the lecturer!' I agreed and apologised for keeping them waiting. 'Don't worry', they said, 'these things never start on time but we like to come early to get the best seats!' I made my excuses and retreated to the dining room, where the dinner was about to be served. At 10 we finished our meal and went back into the bar for a 'little lubrication' and at 10.30 the platform party was duly mustered and processed into the conference room. I was surprised but pleased to see the nuns were still there. Scouts had been duly sent round the other three or four bars in the hotel to round up the audience, who, unknown to me, had also been at the hotel for the last three hours, enjoying the Guinness and the Irish whiskey. This was one of my toughest lecturing assignments and I decided not to show any of the slides I had brought along for the occasion, since with the lights out, I would never be able to tell whether I was successful in keeping my audience awake.

I was very pleasantly surprised when, a couple of years after my move to BOCM Silcock, Professor John Bowman invited me to Reading and asked whether I would be interested in accepting a Visiting Chair in Agriculture at the University. I had already held a Visiting Industrial Lectureship at Nottingham University during my time at Colworth, but it was unusual for agricultural scientists to be offered University Visiting Professorships. I asked about the duties and John said he had heard I had attended a short-course run by the Harvard Business School and that he would like me to give a series of lectures on research management to his post-graduate students and staff. It seemed an attractive offer, so I accepted and for about ten years greatly enjoyed my association with Reading University. Much to my surprise, Ken Durham, the Chairman of BOCM Silcock, was even more enthusiastic about this offer, seeming to regard it as a sign of recognition that the Company enjoyed the respect and recognition of academe. I did not tell him that John Bowman had made it very clear that the position was entirely personal to myself and was nothing whatsoever to do with the company that employed me.

The trade association of the 'agricultural supply industries' was a body known as the United Kingdom Agricultural Supply Trades Association (UKASTA). All three Unilever feed Companies were active members of this Association, and most of the Company Chairmen became UKASTA Presidents.

The Council of UKASTA was comprised of Company Directors. It dealt with 'political' issues affecting the trade, originally requiring discussion with senior civil servants within the Ministry of Agriculture, Fisheries and Food (MAFF), but later also with the European Commission in Brussels. As time went on, many of these topics became highly complex, with legal and scientific implications. The laws governing world trade in agricultural commodities were so numerous and bureaucratic that a strong Legal Committee was set up to read through the small print of new legislation. At the same time, many of the technicalities became more scientific, requiring an appropriate team of applied scientists to be on the look out for errors in the small print which, undetected, could make trade either very difficult or impossible. Thus a Scientific Committee was formed to work in liaison with the Legal Committee.

For over ten years I was appointed by the Council to chair the Scientific Committee and to make regular reports about the practical implications of new laws being placed on the British Statute Book, mainly as a result of new Regulations emanating from Europe. This job was time-consuming and required many meetings to get nonsenses sorted out before damage was done. A good example of how things could go wrong was in connection with a new law which sought to place an upper limit on the amount of Vitamin D3 incorporated into turkey diets, since Vitamin D3, like any other nutrient, can be toxic if supplied in excess of need. This upper limit was based on some old scientific papers written when turkeys grew at about one third of the rate exhibited by modern breeds. If fed below the proposed new maximum allowed for Vitamin D3, all modern turkey flocks would die of acute clinical rickets. The Scientific Committee had to amass a large dossier of information from turkey experiments in order to persuade the 'men at the Ministry' to change their minds. The final weapon in our armoury was to threaten to make public the fact that turkeys would be subject to 'pain and distress' if fed insufficient Vitamin D3, and that their owners could be prosecuted by the same 'men from the Ministry' under the provisions of the *Cruelty to Animals Act*. After much hard work, the law was changed before too much damage had been inflicted on the unsuspecting turkeys.

Another battle which we had to fight with both Whitehall and Brussels was to retain the right to use small amounts of copper in the diets of pig feed. The discovery that copper was an effective, and very low-cost, growth promoter was made by Dr Braude, a scientist in the Pig Department at Reading. Phil observed that when pigs were housed in sties with copper water pipes, they invariably sucked or chewed the copper piping, so that it was always bright. Animals are canny creatures, and often know what minerals are good for them and they seek to replace any limiting trace elements by getting these essential nutrients from any source available. This led Phil to experiment by augmenting pig diets with different amounts of copper salts, and recording the results. It became clear that copper up to inclusion rates of 120 ppm was very beneficial for pigs; below or above that figure their growth and development was sub-optimal.

Based on this work, most well informed pig feed manufacturers added copper to their diets as a means of assisting growth, but when we joined the EEC, copper had to be declared as a 'Growth Promoter' and before this could be done, the manufacturer had to obtain a 'Product Licence' from MAFF to make copper inclusion legal. Now most growth promoters were

34. Group of Guest Speakers at Smith Kline French Pharmaceutical Conference, Windsor, 1981. Author, third from right.

patented and licensed by the pharmaceutical companies who manufactured them. But no pharmaceutical company manufactured copper, a freely available chemical commodity, so it was necessary for someone to submit the data to MAFF and obtain the necessary Product Licence. All the companies in the feed trade needed this licence, so the Scientific Committee of undertook to produce the required dossier of scientific data and submit it to MAFF for approval, CAFMNA subsequently becoming the 'Licence Holder'.

A great deal of effort was put into obtaining the Product Licence for copper by the Scientific Committee and we published this during the course of our work, (89, 98, 100). During my thirteen years as Chairman of the Scientific Committee, a great deal of similar work was conducted, most unusually, by teams of scientists from competing companies (58, 63, 108, 112, 132). In addition to this conjoint work, I published quite a number of papers in various trade and scientific journals, which were judged to be helpful to the trade at large rather than written solely for the indirect benefit of BOCM Silcock (43, 54, 56, 72, 73, 79, 82, 84, 86, 87, 93, 94, 97, 102, 104, 106, 107, 120, 121, 127, 134, 140, 157, 158, 160, 162).

One pleasing aspect of my years as leader of the Scientific Committee was the complete lack of jealousy displayed by the members, who came from competing companies. In their committee work, they worked happily together as members of a team for the public good. As soon as they left the trade association offices, they were at each others' throats, pushing their particular company's products and attacking the opposition whenever opportunity allowed. When it came to putting together a delegation to argue our case with a British Minister or with a Commissioner in Brussels, we arranged for the best people to represent us. I was greatly touched when I left BOCM Silcock and handed over the chairmanship to a friend and colleague from Dalgety, Dr Brian Cooke. On this occasion I was invited to a dinner given in my honour and presented with a complete outfit of academic dress in order to 'remember my Scientific Committee friends when I was on parade in Edinburgh'. I have worn the academicals with pride for the last sixteen years.

A daunting task which I had to deal with as part of my Committee duties was to inform the farming public about the acceptability of what, to them, seemed strange and unusual raw materials (50, 56, 60, 97). Most of these were imported by-products from tropical countries and indeed the animal feed trade originated from the use of some of these commodities, such as the various 'cakes' produced as a result of extracting oil from oily fruits like coconut and oil palm. The fruits were imported whole for expelling the oil as a raw material for the manufacture of soaps and margarine and the resultant crushed seed, known as 'cake', was then used as a fuel for heating the steam boilers which powered the factory operation. It was noticed that horses used for transport into and out of the crushing plants were partial to eating this cake and that it gave their coats a glossy 'bloom', which was regarded as a sign of good health. The factory managers therefore thought they could make a useful sideline to the business by selling the cake to farmers rather than burning it in the furnaces, and this gave rise to the compound feed trade and to the term 'cattle cake', which is still in use today.

The feed trade was always on the lookout for novel and economic raw materials. The UK cereal crop totals about 22 million tons, and for each ton of grain another ton of straw is produced. Traditionally, most of this straw was burnt in the field, but the practice had become increasingly environmentally unfriendly and eventually in England (although not in Scotland) laws were passed to ban straw burning. Straw is, potentially, a source of nutrients but most of these are encapsulated within woody cell walls so that the cell contents are unavailable to animals eating the straw. However, if these hard cell walls could be broken down, the nutrients contained therein could be released.

Whilst I was at Colworth a lot of work was done on methods of breaking down these cell walls.(52, 61, 62, 66, 68, 69, 71, 74, 83, 90, 99, 115, 117, 125, 130). One of the most efficient processes was the use of sodium hydroxide (NaOH); a powerful alkali which attacks the lignin but does not damage the more nutritious cell contents. This work is tricky to carry out on the farm as NaOH is a dangerous chemical, which has to be handled with care. Furthermore, the concentration of the alkali is critical – too little or too much are equally unhelpful and so it is essential to make sure that the correct concentration of alkali is used.

After about ten years the problem was solved and a new process, leading to a new product, 'Nutritionally Improved Straw' (NIS), was ready for launching. For the safety considerations, the process is best carried out in purpose-built factories. In the early 1970s proposals were accepted by the BOCM Silcock Board to go ahead with the erection of a pilot plant. However, Ken Durham foresaw that the production of NIS would escalate, and to fend off the opposition, he thought that five NIS plants should be built to gain a flying start. Such a large capital expenditure had to go to the Unilever Main Board for authorisation and Ken Durham went to London to argue the case for the money. He was convinced of the strength of the argument, but worried that the non-technical Main Board members might not understand what the project was all about. As I had been involved in conducting experiments on the process and on the feeding of NIS to cattle, he asked for my advice on how to sell the idea. 'I only have about three minutes', he said, 'so the story has got to be simple and powerful'. I asked him if he had heard the analogy of the elephant with the can of baked beans. He replied in the negative and asked me to explain. 'Consider feeding unopened cans of baked beans to elephants', I said. 'The elephant could swallow the cans, but they would come out of the other end intact, except that the paper wrappings would have been digested. Now if you made just a few holes in the cans with a tin opener, they would then go through the elephant but the contents would be digested on their passage through the gut. Eventually the tins would pass out of the elephant clean and empty. So it is with NIS. The NaOH treatment will punch holes in the woody cell walls so that the cattle can digest the contents. All that will pass through the animals will be the undigested woody fragments of the cell walls'.

I assume that the trick worked, for BOCM Silcock got its money and the NIS plants were built and duly produced pelleted NIS nuts. Unfortunately, the economics of the operation eventually turned sour. The price of competing raw materials, including cereal grains, came down and the NIS product, which was relatively expensive to produce, priced itself out of the market. However, the basic idea of treating fibrous materials with NaOH is still used in various parts of the world, although mainly as a low cost on-farm process rather than as a safer and more expensive factory-made product.

One cannot work for a major commercial company for very long without realising the overriding importance of economics and marketing as pre-requisites of successful business management. The same is true of farming, and during my Unilever period I encouraged my team to be watchful for new innovations in these areas. Although my colleagues were mainly concerned with animal nutrition rather than other subjects, we did publish work in this vital field (78, 91, 133). I was fortunate in having very numerate colleagues with good computer skills: David Filmer, the Chief Nutritionist, and Paul Roberts, the Chief Statistician, being widely recognised as pre-eminent practitioners of these important arts. Together we introduced several new computerised management systems. One, known in the Company as 'P360' (Project run on a 360 computer), was designed to put greater precision into the individual feeding of cows whose milk yields were not routinely recorded (53). Another, using the results of work carried out by Professor Jack Payne and his colleagues at the Compton Animal Research Institute, was a programme based on blood-testing a sample of cows in a milking herd at different stages of their lactation for the presence of essential metabolites essential for good health and high performance. This technique was known as 'Metabolic Profiling' and our company successfully introduced this programme for sale to interested customers who wished to avoid their high-yielding cows suffering from the debilitating so-called 'Metabolic Diseases' (64, 88).

During my time with BOCMS I was invited by the Agricultural Research Council to serve on a committee chaired by Sir Kenneth Blaxter, then Director of the Rowettt Research Institute, which was looking into new methods of measuring the Energy and Protein requirements of cattle. This work, which spanned several years, resulted in the introduction of a new Metabolisable Energy system and, later, a new Metabolisable Protein System. As the trade representative on Ken's Committee, I was responsible, together with my colleague Geoffrey Alderman from ADAS, for helping to introduce these new concepts to the feed trade and thence to technically-minded livestock farmers. This was interesting work, done in collaboration with the leading animal nutritionists at that time, and led to the publication of several papers (81, 116) and eventually to the production of a textbook aimed at agricultural advisers and *avant garde* cattle farmers (126).

Another topic which attracted my attention was the subject of Energy, the different efficiencies of producing energy from different sources and the economics of the exploitation of energy at the farm level (105, 122). This subject was not capable of the usual on-farm experimentation but nevertheless the results of these theoretical studies were of importance in agricultural practice, and this area of study has expanded considerably since our early attempts to grapple with it. New experimental designs, suitable for application to herds or flocks of animals, also attracted our attention and once again the expertise of Paul Roberts and David Filmer was an immense advantage to our team (57, 137).

But throughout the 1970s and early 1980s most of our work was connected with straightforward matters of animal nutrition. Many of the studies were naturally confidential and only saw their way into BOCMS Development Reports but, as at Colworth, I encouraged open publication of those parts of this work which were not commercially sensitive. These studies covered calves (96, 105, 118, 132); cattle (95, 119) and sheep (109). I also developed an interest in the subject of the many different systems of beef management and a few papers were published in this field (77, 128, 139). Finally, remembering John Hammond's advice to me when I went to work in Uganda, I endeavoured to interest the other Unilever businesses in the exciting things going on in the animal production world and I published a 'popular' paper on animal fertility in the *Unilever House Journal* (80), believing that 'fertility' might be of more interest to the lay reader than 'nutrition'!

Although I was enjoying my interesting and varied work with BOCM Silcock, changes were afoot. The Research and Development budgets were being either frozen or cut. My small but efficient band of headquarters staff were becoming increasingly difficult to replace as folk left, on promotion, to go to other organisations. BOCM Silcock Chairmen came and went, and at each change-over, policies altered and priorities shifted. I had increasing difficulty in fighting my corner and was wondering how long it would be before another merger or take-over materialised. However, before there was any need to do anything drastic, an offer came out of the blue to go back to Edinburgh University and 'Meet the Chair Committee' without any obligation. Apparently my name had been put forward as a possible successor to Professor Noel Robertson, who was retiring from the Chair of Agriculture and Rural Economy. I packed a suitcase, flew up to Edinburgh, was duly interviewed and that same evening was offered the job by Dr John Burnett, the Principal and Vice Chancellor of Edinburgh University.

Back at Basingstoke, I handed in my resignation and was told I was an idiot and would soon regret my stupid mistake. To my disgust, pressure was put on my wife with an offer to double my salary and offer me a seat on the board in order to get me to withdraw my decision to leave. However, as

Bunny pointed out, I neither would nor could go back on my word. I accepted the Chair and left after working for more than my statutory three-months' period of notice. Unknown to me at the time, plans were afoot to sell the whole of Unilever's interests in animal feed manufacture and its associated activities and the business was eventually acquired by Pauls of Norfolk. The research and development farms were sold, most of the high-calibre scientific staff were transferred or made redundant and an important chapter in the books of post-war British Agriculture was closed. The take-over of BOCM Silcock by Pauls was not the only bit of restructuring which occurred at that time. The number of feed companies has halved since I left and many well-known and respected names in the farming world have disappeared without trace. I find this sad but inevitable. I am pleased that I left when I did and was able to complete the last stage of my career back in Edinburgh.

However, even in academe, the drive for yet more economic efficiency is impelling Colleges and Universities into the equivalent of the mergers and take-overs in the commercial world. My *alma mater*, Wye College, is to merge

35. Allan Price, Chairman, BOCM Silcock, reading letter from the Chairman of Unilever Ltd., Mr Kenneth Durham, at the farewell party to mark the author's move to Edinburgh, December, 1983.

with Imperial College, London the year this book is being written and an unique and successful institution will inevitably lose some of its former autonomy and unrivalled academic charm. The existence of the six excellent veterinary schools in the UK was seriously challenged by the 'Riley Committee' a decade ago, and I have no doubt that the same pressures to 'economise', by reducing the admittedly very high costs of veterinary education, will return to bedevil us again in the future. And finally, as we will see in a later chapter, the School of Agriculture at Edinburgh University is currently announcing its intention to shed general agricultural degrees from its curriculum. The reason is the decline in the number of students opting for agricultural courses and the consequential high cost per student place. It would seem that mergers and take-overs lie around every corner and we are moving away from the principle of Small is Beautiful to a new imperative that Biggest is Best. At times I am grateful that I retired before these movements came to the fore and thus was able to pursue my own career at a time when institutions, both in and outside the public sector, were more stable than they are today.

CHAPTER 12

Return to Academe – Edinburgh Revisited

'Breathes there the man with soul so dead
Who never to himself hath said
"This is my own, my native land!"
Whose heart hath ne're within him burn'd
As home his footsteps he hath turn'd'

Sir Walter Scott

WE ARRIVED BACK IN Scotland on a cold dreich winter's day in January 1984. The furniture removal vans which had brought our belongings up from Farnham got stuck in a snow drift in the driveway to our new home. It was a somewhat cold home-coming!

In those days the Principal of the Agricultural College had an official residence at Bush on one of the College farms. Noel Robertson, my predecessor in the dual capacities of Professor of Agriculture and Principal of the College, had lived there very happily for fourteen years and it was naturally expected that we would do likewise. However, although we had lived on farms for most of our fourteen years overseas, there were clearly disadvantages, particularly in times of high inflation. Retirement was not all that far away and the tied cottage of the Principal's house could not be taken with one into retirement. It therefore did not take us long to decide that we would have to leave the Principal's official residence and buy a home in or near Edinburgh. Bunny spent a lot of time diligently house-hunting and six months later we moved into the nineteenth house since our marriage. In those nineteen moves we had journeyed thousands of miles but ended up about one mile from where we started.

The post in Edinburgh was a combined one, linking the East of Scotland College of Agriculture to the Department of Agriculture in the University. I was both Principal of the College and a Professor in the Faculty of Science and the dual function carried the title 'Head of the Edinburgh School of Agriculture'. The two jobs and their titles were not unique. A similar arrangement applied to Aberdeen, the equivalent College there being the North of Scotland College of Agriculture. However, Edinburgh had the distinction of having the oldest Chair of Agriculture in the English-speaking

36. The Board of Governors of the East of Scotland College of Agriculture, 1984. Seated, from left, Helen Scott, Secretary to Reg Boyd; Reg Boyd, Secretary to the Board of Governors; Professor Charles Kemball, FRS, FRSE, governor; Dr Jean Balfour, CBE, FRSE, governor; Mr Alex Barbour CBE, Chairman; the author; Mr Alexander, governor; Mrs Ann Syme, administrative officer.

world. The Chair was founded in 1790 by a wealthy landowner, William Johnstone Pulteney, and the first occupant was a certain Andrew Coventry, a medical graduate of Edinburgh University, who occupied the post for forty years (171). It is of great interest to many Scots that there was talk at the time of the first appointment of the Chair possibly having been offered to Robert Burns, Scotland's national bard. He was in Edinburgh when the post was advertised and it was brought to his attention by Mrs Frances Dunlop, one of the well-known Edinburgh society hostesses at that time. However, Rabbie did not apply for the position. In a letter to Frances Dunlop, dated 7th July, 1789, he wrote, '*I have no romantic notions of independency of spirit, I am truly obliged to you and Dr Moore for mentioning me to Mr Pulteney. From the manner in which God has divided the good things of this life, it is evident that He meant one part of Mankind to be the Benefactors and the other to be the Benefacted; and as he has thrown me among the latter Class, I would wish to acquiesce with cheerfulness. The professorship is I know to me an unattainable object . . .*'

I had the privilege of becoming the ninth holder of Pulteney's Chair, which, to give it the correct title, was the Chair of Agriculture and Rural

Economy (see Table 2). Thus in two hundred years, the average tenure of the Chair was over twenty years, which is remarkably long by present-day standards. As I only occupied it for six years, I rather feel that I let the side down. However my successor, Professor Colin Whittemore, has now held the post for over ten years, so the length of tenure is now improving. Sadly, though, there is currently talk of closing down agricultural teaching in the University of Edinburgh, so it may be that Colin Whittemore will be the last occupant of this ancient Chair, which, if it comes to pass, will be very sad.

Table 2
Holders of the Chair of Agriculture and Rural Economy

William Pulteney	1790–1830
David Low	1831–1854
John Wilson	1854–1885
Robert Wallace	1885–1922
James Scott Watson	1922–1925
Ernest Shearer	1926–1944*
Stephen Watson	1944–1968*
Noel Robertson	1968–1983*
Peter Wilson	1984–1990*
Colin Whittemore	1990–

As the Chair was founded in 1790, it followed that I had the privilege of being in post when the bicentenary of its foundation was celebrated. The year was a wonderful one, with a major event of some form or other taking place every month (168, 172, 173)., The most splendid function was a superb dinner organised in the Fellows' Dining Room at the Highland Show Ground at Ingliston. Prior to this event, we undertook some research to find whether Edinburgh could lay claim to the oldest Chair of Agriculture in the world and were intrigued to find that this distinction went to Italy, where the first Professor of Agriculture had been appointed at the University of Padova in 1764. I therefore sent an invitation to the current occupant of the Italian Chair to attend our celebration dinner as our guest and was a little miffed to find that, for several weeks, I got no reply. Eventually an answer arrived. The reason for the delay was that Padova had not realised they had the distinction of being the first in the field, so it had taken them some time to verify their claim to fame. Apparently, their own Bicentenary had passed unnoticed and uncelebrated in 1964.

*These four holders of the Chair were also Principals of the East of Scotland College of Agriculture during their period of office.

When I moved back to Edinburgh in 1984, the Chair was firmly rooted in the Faculty of Agriculture and the Department awarded degrees in Agriculture, Animal Science, Plant Science and Agricultural Economics. Most students studied for the four-year Honours degree, but the distribution of students between the four different sub-disciplines was unbalanced. Only very small numbers studied for the Plant Science and Agricultural Economics options. Even in the two popular branches of the subject, the student numbers, although reasonable, were steadily declining.

The Faculty then had two sister Departments to Agriculture – Animal Genetics and Forestry – and both these specialised disciplines suffered from the same problem as agriculture: declining student numbers. In fact, the Genetics Department (which I had entered in 1949 as a postgraduate student) was moving away from undergraduate teaching to postgraduate studies, leading to postgraduate diplomas (similar to the one I had obtained in 1950) and M.Sc. degrees. The Forestry Department had balanced the decline in student numbers studying forestry by offering options in ecology and wild-life management. This initiative worked well, with rising numbers of students studying for these alternative courses.

This wane in the popularity of agriculture concerned me. My overseas career had taught me that the world was getting steadily shorter of vital food supplies and that the global need was for more, and not fewer, trained agricultural scientists. It would have seemed a good idea to have marketed the excellent agricultural courses at Edinburgh to overseas students. Unfortunately, however, previous UK governments had increased the costs of British University education to foreign students, so that they were forced to pay 'economic fees', with no government support or subsidy This short-sighted policy inevitably drove prospective clients to Canada, America, Australia and other developed countries, who continued to give generous support to such deserving students. In addition to this, I had, in my own small way, been responsible for helping to get higher education going in the previous British colonies. It was often argued that students from African countries, where I had worked, would be well advised to take their first degrees in their home countries, where the knowledge acquired would be directly applicable to local conditions. But however one argued the case for 'local' versus 'overseas' higher education, there was no way that Edinburgh could solve its logistical problems by robust recruitment overseas, except at the post-graduate level, where such financial considerations did not apply to the same extent.

Whilst these pressures were developing, the small Agriculture Department had only one Professor – me. I could give some leadership in animal science, but I needed help and support by experienced people with a good record in the other subject areas. My predecessor, Professor Noel Robertson, had realised that he needed two additional Professors to assist him and he was successful in persuading the University to create senior posts in both Animal and Plant Science. Unfortunately, the Crop Chair became vacant about a year before I arrived, when the holder, Graham Milbourn, moved to Cambridge to become head of the National Institute for Agricultural Botany. Then, almost on the day I assumed office, the Professor of Animal Science, John Prescott, informed me that he had been offered the Directorship of the Grassland Research Institute at Hurley and would be moving south. So my first year was spent persuading the Dean of Science and the Principal of the University to fund these two vacant Chairs. Both these posts attracted a strong field of applicants and in the event both successful candidates were internal members of staff. The Animal Science Chair went to Colin Whittemore, previous Head of the Animal Production and Advisory Department in the College, while the Crop Science Chair went to John Holmes, on promotion from his previous position as Head of the College Crop Production and Advisory Department.

However, because I was concerned about the lack of leadership in agricultural economics, I tried to go one step further and create a new Chair in the rapidly developing subject of 'Agricultural Resource Management'. Both the Dean and the Principal thought that this proposal was one step too far. It was even hinted that I was developing a reputation for 'Empire Building', and it was made clear that, if I really wanted to create a new Chair in Agricultural Resource Management, I would have to find the necessary funding from non-University sources. It seemed to me that some of the obvious potential benefactors of the new Chair were the banks, but apparently the resources required to endow a Chair were beyond the means even of any single Scottish clearing bank. The going rate for a fully endowed Chair was then over £1m. I therefore approached the Scottish clearing banks jointly, through the aegis of the Scottish Institute of Bankers, and was eventually successful in providing the necessary pump-priming to finance the new Chair for its first five years. Although no guarantee was given, I assumed that if we recruited a fitting occupant for the new Chair, and if he remained in post for at least this five-year period, the University would then accept the continuing responsibility for its future funding. So the Chair was duly advertised and amongst a strong list of candidates, an outstanding application was received from Barry Dent, then occupying a Chair in a similar subject at Lincoln College, New Zealand. Barry was duly appointed and did a great job in building up a new and strong research department in this all-important area.

By the end of 1985, therefore, I had a full team of three Agricultural Professors supporting me, on whom I could rely to lead research and teaching in their respective areas and to bring in the necessary money to enable the research base to be expanded. Because of the close inter-relationship of the East of Scotland College and the University, all three Professors also acted as Heads of the appropriate conjoint Division, comprising both College and University staff, so the resources at their disposal were far greater than would have been the case had they undertaken normal University duties only.

Although money, as always, was the limiting factor in building up my Professorial team, I had great support from the various Deans of the Faculty of Science who were in office during my time. The first was Mike Yeoman, a botanist and as such, someone who fully understood the need to enhance the status of Plant Science within the Agricultural Department. Mike was one of my close colleagues, who nominated me for Fellowship of the Royal Society of Edinburgh and Mike and his wife Erica became good friends. Mike was followed by Evelyn Ebsworth, a structural chemist, and I knew Evelyn well both on and off the campus as he was a member of our Episcopalian Church, Christ Church, Morningside. In addition to being a chemist of international repute, Evelyn was also an accomplished musician and he and his wife had good singing voices. Tragically, Evelyn's wife died, but Evelyn married again, and shortly afterwards left Edinburgh to be Vice Chancellor of Durham University. Evelyn was followed for a short period by Murdo Mitchison, son of the well-known writer, Naomi Mitchison. Murdo was a zoologist and had been at Edinburgh as a contemporary of C.H. Waddington in the early post-War years. Murdo was succeeded by John Mavor, an electrical engineer. John was a very able administrator and went from Edinburgh University to become Principal and Vice Chancellor of the new Napier University, which, when I returned to Edinburgh in 1984, was just a Technical College. Like me, John had spent part of his career in industry, with an electrical engineering company in Fife, and we both fully understood the need to bring a more business-like approach into the administration of the University. John also was a firm supporter of Burns Suppers, and was our guest of honour when we celebrated the Bicentenary of the Chair of Agriculture at a magnificent Burns Supper-cum-Celebration Dinner at Borthwick Castle in 1990.

There were five other Divisions within the Combined School of Agriculture. In Noel Robertson's time, the University Department of Microbiology had been incorporated into the School of Agriculture, and this was headed by a full University Professor, John Wilkinson. I inherited this structure, which, although unusual, worked well. The College had two other large Departments, deployed over the whole of the East of Scotland. One was an Advisory Department, under the leadership of Ken Runcie, and the other was an equivalent Veterinary Division, working from three outlying veterinary investigation centres, headed by Ogilvie Mathieson. There was also a Farms Division, led by Murray Black, the Farms Director, and an Administration Division, headed by the Secretary to the School of Agriculture, Reg Boyd. Reg had previously worked for our paymasters, DAFS, and had a good inside knowledge of the way the civil service worked. He was due to retire in a year's time and so did not have to cope with the radical changes that lay ahead. His eventual replacement, Stuart Ainslie, a graduate chartered accountant, was an excellent financial administrator and he became a very highly regarded member of our senior management team. Nothing was too much trouble for Stuart. At Divisional Managers' meetings he would say, 'If you need more and better management information, let me know and I will provide it.' He delivered on his promises and as a result, I and my Divisional Managers had a firm control over the monthly College finances. Although they were more complex, they were by far and away better managed than the rather traditional 'historical' accounts, which were all that could be produced by the University Finance Office at that time.

The staff and facilities under my control were therefore extensive. Administration was complex. When wearing my College hat, I was responsible to my Board of Governors of the East of Scotland College of Agriculture and also to my paymasters, the Department of Agriculture and Fisheries for Scotland (DAFS) within the Scottish Office. On the other hand, wearing my University hat, I was responsible to the Dean of the Faculty of Science, and, through him, to the Principal.

The *modus operandi* of this complex structure was inherited in the main from my predecessor. He had created the new Divisions and he ran much of the School through regular meetings of the Divisional Managers. I continued this arrangement, regarding the Heads of Divisions as the senior members of my team, but I changed the pattern in respect of the University staff. Noel had appointed, or caused to be elected, Professor John Prescott as the Head of the University Department of Agriculture. I did not like this solution, as it tended to ear-mark the Head of the School as a 'College man', who regarded his University duties as being of lesser importance than his College responsibilities. I therefore made it plain that I wished to be the Head of the University Department and as such to have direct access to the Dean and Principal and regular day-to-day contact with the University Agricultural staff.

I continued this arrangement until my final year, when I suggested that the University staff should elect another Head of Department, to give continuity to University affairs after I had left. They duly elected Professor Colin Whittemore and, as indicated above, Colin in due course succeeded me in the Chair of Agriculture and Rural Economy, going on to become Head of the Institute of Ecology and Resource Management (IERM) when Agriculture and Forestry were put together. I regretted, as did Colin and our colleagues in Forestry, the dropping of the key words 'Agriculture' and 'Forestry', but this was in line with a well established trend, since these key words had become politically unacceptable. This was because both Agriculture and Forestry were increasingly being regarded as branches of Applied Ecology and Biotechnology. The reason was twofold. Overproduction of food in the UK meant that the government policy of 'Food from our own Resources' was no longer valid and secondly, the various 'Green Lobbies' were increasingly critical about such subjects as the over-use of fertilisers and pesticides and the alleged cruelty implicit in intensive livestock farming. I fear that, taken to its illogical conclusion, this will eventually mean that all the teaching of the two important subjects of agriculture and forestry within the major British Universities will be conducted by so-called 'pure' scientists and economists of one sort or another. Few of them will have detailed knowledge of the facts of farming and, especially in forestry, they will have increasing difficulty in seeing the wood for the trees. I dread to think what Ken Smith, my early farming mentor, would say about this. Perhaps he would have remarked, 'I told you so!'

The scientific staff, in both the College and the University, regarded as their prime task teaching their students to the best of their ability. They saw their second task as making themselves useful to the agricultural community by giving free advice and conducting experimental programmes to reinforce their teaching and advisory activities. In 1984, not only was all advice to farmers provided free, but the research programmes underpinning it were supported by grants from the government, DAFS, or by parastatal organisations, such as the Marketing Boards. In addition, the farms controlled by the School of Agriculture were profitable and hence able to support a fair amount of relevant R&D without reliance on outside funding. I knew that at Reading University, where I had been a Visiting Professor, some 20 per cent of the Agriculture Department's budget came from third-party sources. In 1984 a few Edinburgh staff were in receipt of modest funding in this way, but this was the exception rather than the rule. At one of the first University Agriculture Department meetings I presided over I said I was concerned that only about 5 per cent of the budget was derived from third-party funding, and expressed my opinion that this had to rise. I was greeted in stony silence. Did I not realise, I was reminded, that this would mean a commercial bias being put on research programmes? Farmers would not be able to pay for R&D and thus the money could only come from commercial agricultural organisations and therefore would have strings attached. I mentioned that I had come directly from such an industrial background and, provided negotiations were properly conducted and contracts drawn up accordingly, I could not see that income derived in this way was tainted. I argued that commercial money was not the only means of supporting a more vigorous R&D programme, pointing out that the Research Councils and major charities also supported high quality research. This was countered by the argument that such moneys normally went to 'pure' science projects, rather than to the more applied and strategic work which was the main thrust of the Agricultural Department's activities.

Fortunately some of my more senior colleagues could see clearly from which direction the future winds would blow. They wholeheartedly supported moves to increase our third-party funding, but there remained a sizeable rump of staff who clearly disliked the messages they were receiving and wished to maintain the *status quo* in order, as they saw it, to retain their independence. With the help and encouragement of my team of new Professors, we slowly applied pressure so as to garner increased support for more ambitious R&D programmes and over the course of the next few years moderate success was achieved in this area.

The other matter that concerned me was the relatively modest publication record of many of the University staff. I pointed out that our colleagues in the College were producing, on average, more papers in refereed scientific journals than the University staff. I thought this strange and a matter which needed attention. The response was that the College staff were employed in order to do teaching and research, or advisory work and research, whereas the University staff were originally recruited mainly to teach. Some of the University staff clung to this viewpoint throughout my period of office, but happily the majority responded. Those who did first-rate research (and there were many in this category) were glad to see encouragement being given to write up their work for the learned journals. Others saw the need for change and, over time, produced good publication rates, but the going at times was tough and I must have been regarded as difficult. Strangely, there was less of a problem in pointing the College staff in this required direction. Being employed with contracts that spelt out the need for getting the teaching, advisory and research acts closely together meant they could see it was in their interest to co-operate in all three areas. However, there was a noticeable

37. Receiving Russian Scientific Delegation in the author's office at the School of Agriculture, Edinburgh, 1987.

tendency to do good research but then put off the task of writing it up for publication.

As in my Unilever days, I thought the best way of encouraging better publication rates was to lead by example. I was no longer in a position to engage in experimental work; to do so would have encroached on the domains of my Professorial staff and been resented. I could, however, embark on the production of more review-style articles and papers dealing with possible future strategies for agriculturally related research (146, 154, 160, 161, 163, 166, 169). I responded to a request from the Royal Society of Chemistry to give a lecture on the 'Role of Fats in Animal Feeds' (141), and I agreed to write an article on the interaction between farms, woods and forests as part of the plans to bring the Agriculture and Forestry Departments of the University closer together (152). I was invited by the Royal Society of London to give a paper on the agricultural aspects of 'Global Warming' (167) and this led me to take an increasing interest in the whole subject of environmental sustainability (180). Later on, when I became a member of the University Grants Committee and subsequently of the Research Assessment Exercise (see chapter 13), I was thankful that I had insisted that published research was as important as good teaching. During my time at Edinburgh, regular publications have rightly become a requirement, not an optional extra, for academics.

Colin Whittemore and Barry Dent accepted the challenge of getting research work published in the top journals. Thus the number of publications in their respective areas of work increased enormously under their encouraging leadership. Because of the close integration of College and University for most of my time, it was not unusual for colleagues employed by the College to be more productive with papers for publication than members of the University staff. Keith Smith (Soil Science), Geoff Simm (Animal Breeding), Carol Duffus (Plant Biochemistry) and John Oldham (Animal Nutrition) were all College employees, but their publication record was outstanding and could be favourably compared with that of comparable staff from any leading University Agricultural Departments in the UK. On the other hand, some of my academic colleagues were exceedingly good with students, giving unstintingly of their time to their supervision, helping with academic administration in the Faculty Office and assisting in the all-important task of recruiting future students into the Department. Outstanding among them were John Manson, Charlie Hinks, Ron Wilson and Alastair Stott on the University pay-roll; and Joe Lennard, John Percival, Linda Chadwick and Alison Pake on the College staff. All these colleagues, and many others as well, took an interest in the students after they had left Edinburgh, playing a leading part in the organisation of reunions and other events for the former students, both University and College alike.

Unfortunately the new and increasing emphasis on the output of good publications does lead to some staff thinking that teaching is now only second in importance to research. This is unfortunate, but the signals being given by such bodies as the University Funding Councils and their regular Research Assessment Exercises do not help new staff to appreciate that it is the correct balance between teaching and research which is needed, not an over-emphasis on research alone. I fear that younger and newer staff may fall into this trap of forgetting the main mission of the Universities. Indeed, a recent recruit to the Department (who joined after I had left) was said never to be available to his students except by advance appointment, conducting most of his research work at home and only coming into the Department to give the occasional lecture.

The University administrative procedures were familiar to me, as they were similar to those at all the other Universities with which I had been connected. On the other hand, the College had an unique system of management which needed a lot of getting used to, and was eventually forced to undergo dramatic change. From my viewpoint, I was subordinate to a Board of Governors, most of whom had been appointed by organisations such as the funding body, DAFS, the Farmers' Union, Local

Councils and other representative bodies. The Chairman was Alec Barbour, factor to the Duke of Atholl, who earlier in his career had had experience of working for the College as a local agricultural adviser. He was a shrewd and conscientious Chairman of his mixed Board and I had an excellent working relationship with him. Bunny and I became good friends of Alec and Hazel, his wife. I was not a full member of the Board of Governors, but 'in attendance', which I found somewhat unusual. However, since this did not impair our working relationships in any way, I was content with the arrangement. I had only one difference of opinion with my Board, and that was over the matter of a Corporate Plan for the College. I was concerned that we were lacking a clear sense of mission and that most of the decisions were being made, not by the Board, but by our paymasters, DAFS. I thought we would be better placed if we became more proactive and for this reason I spent a lot of time with the Divisional managers in devising a five-year Corporate Plan. When this was eventually presented to the Board for their approval, they were somewhat uninterested. Since the DAFS had not called for such a plan, they reasoned, why had we bothered to draw one up? This disappointed me, but I had to accept that I had spent insufficient time in advancing the idea of the plan and therefore was partly to blame for the indifferent way in which it had been received. The plan was useful for motivating the staff and giving them a clear goal, but it was never backed by formal acceptance by my Governors.

On another matter I was very fully supported by the Board. As already mentioned, a lot of the experimental work of the College was carried out on its farms, much of it being paid for out of farm profits. Fortunately throughout my time the farms were profitable overall, although some areas were financially more successful than others and consequently bore a larger share of the costs. As more detailed research was taking place on the College farms, the costs were rising alarmingly and it was clear that the time would come when the farms lost money, not because of unsuccessful commercial operations, but because the burden of the R&D programmes was becoming excessive. I therefore introduced a system by which all those conducting R&D on the farms had to fund their work, one way or another. The farms needed their profits to pay for renewals and capital improvements and so could not continue to subsidise R&D programmes over which they had no control. One of the Governors, Mr Sandy Inverarity, was appointed Chairman of the Farms Committee and he was a tower of strength in getting this new policy of 'pay as you go' accepted. It was a long struggle. There was inevitably special pleading that some parts of R&D had greater difficulty in attracting research funding than others and cross-subsidisation was claimed to be essential. However, we persevered with the policy of self-sufficiency and over time were successful in implementing it. The Edinburgh farming operations were, as a consequence, the envy of those responsible for the farms at the other two Colleges, which had much greater difficulty in making ends meet year on year.

But the College Board of Governors and my pay-masters in the DAFS were not the only management structure with which I had to work. The three Colleges, Edinburgh, Aberdeen and Auchincruive (in Ayrshire), were loosely tied together with an organisation known as the Council of the Scottish Agricultural Colleges (COSAC). In 1984, when I first took up my duties, this Council had a Permanent Secretary, a staff member seconded by one of the Colleges. The three College Principals took it in turns to act as chairman of these Council meetings. This was a most difficult administrative structure and not one conducive to arriving at the optimum decisions in an efficient manner. The three Principals, naturally, spent a lot of time fighting their own corner and most of the decisions were really compromises which had insufficient impact on optimising the overall efficiency of the three-College system. In 1986 the first step towards integrating the three Colleges was taken with the formation of the 'Scottish Agricultural College' (SAC) under the Chairmanship of Watson Peat, the former Chairman of Governors of the West College. The new post of Chief Executive was advertised, and Dr Eric Thain was appointed to this position, with the former Secretary of COSAC (Ron Harkess) as it full-time Secretary. The new structure was set up in Perth for purely political reasons, since Perth was in the middle of the triangle of the East, North and West Colleges. In fact, this meant that it was equally inconvenient for all three Colleges (whilst not being convenient for any one of the three) and it also meant that the 'Headquarters' was divorced from all but a few of the staff of the new organisation. This new structure was not well regarded by the staff and was not perceived by the outside world as a major step towards closer integration. In practice, the three Colleges continued to do most of their work relatively independently of each other whilst realising that top policy decisions were now being taken by a small number of people at this somewhat remote location.

The second step in the move towards full integration was taken in 1988, when Lord Sanderson, the Scottish Minister of Agriculture, set up a Committee of Enquiry to look into the future management of the three Scottish Colleges. Lord Sanderson appointed Sir Alwyn Williams, then Principal of Glasgow University, as Chairman of the Committee, which deliberated for a year, finally recommending that the three Colleges should merge into one, with a single Principal and a single Board, but operating on three different sites, East, West and North. DAFS accepted the recommendation and the three Colleges lost their individual autonomies. This recommendation was sensible and worked, but a subsidiary recommendation was, at any rate in my judgement, not so easy to accept. This was that the close ties that bound the College and the University Departments of Agriculture together at Aberdeen and Edinburgh should be severed. The same did not apply to Auchincruive, as Glasgow had closed its Department of Agriculture many years previously. I was concerned that these proposed radical changes were not entirely in the best interests of Scotland and of Scottish agriculture. I could see the wisdom of uniting the three Scottish Colleges more closely together, but for the life of me I could not see the sense of severing the links between the Colleges and their respective Universities, which were the main *raison d'etre* of the 'Scottish System'. I therefore wrote many letters to the Scottish daily and agricultural press, and also published three papers, explaining the System in the hope that if more people realised what it was and what it offered, it would not be dismembered (148, 150, 179). Alas, my efforts were in vain!

I have thought it necessary to spell out in some detail the changes in the administrative arrangements which controlled the destiny of the three Colleges. These discussions took up a great deal of my time – more than I would have wished – and they marked a very major change in the way agricultural institutions in Scotland were run. They also resulted in the eventual demise of the 'Scottish system', in which University and College teaching, R&D and advisory activities were all co-ordinated. The system was not perfect, but it was nevertheless a good arrangement which was the envy of many of our colleagues in England and Wales. The only comparable system is that which still operates in Northern Ireland, where there are still close links between the Professor of Agriculture at Queens University, Belfast and the advisory system run by the Department of Agriculture for Northern Ireland (DANI). That formula remains unchanged, but it never had the complexity of the 'three Colleges in one' system of Scotland.

However, there were many other exciting things going on and life was not all in-fighting and down-sizing. I had served for a few years on the Agricultural Advisory Committee of the British Council in London, and in 1989 I was honoured to receive a letter inviting me to be the leader of a British Council Mission of Deans of Agriculture to Canadian Universities. I duly assembled a team of eleven Deans and in August we set off on a most interesting and instructive tour, starting in Nova Scotia and ending in Vancouver. We were most hospitably received wherever we went and sumptuously entertained by our hosts. As leader of the Mission, I was expected to make a large number of speeches, not simply thanking our hosts for their generous hospitality, but also giving short talks on the British University system and on our initial impressions of Canadian Higher Education. This was quite a marathon, with a new campus being visited every other day, and a great deal of complex travelling. The trip went without a hitch and I was very bucked when one of my team complimented me by saying that I hadn't repeated the same joke once during the whole tour.

A great deal of home entertainment was called for during our time in Edinburgh and Bunny rose to the occasion magnificently. She is an exceptionally good cook and her skills were called into action on numerous occasions. We greatly enjoyed this social side to the work and have kept in close contact with many of the friendships formed in this way. Another activity which we pursued whenever possible, usually at weekends, was touring Scotland and either hill walking or doing a bit of bird watching. We managed to get around most of Scotland, with the exception of some of the smaller outer isles, and were amused to find that, after seven or eight years, we seemed to know more about Scotland than some of our friends and neighbours who had lived there all their lives.

I greatly enjoyed working with the students, although it was difficult to get to know them all as well as I would have wished. I decided to concentrate my teaching on the first year, remembering the advice of Professor Witteridge when I was a student of animal genetics. I gave a series of lectures called 'Principles of Agriculture' to both the Higher Diploma students and the degree students, with the two courses running in parallel. This course was not examined and was primarily aimed at those, mainly in the degree class, who had little or no farming knowledge prior to coming to Edinburgh. I topped this course up with a voluntary half week, touring different parts of Scotland by mini-bus, looking at the diverse pattern of Scottish agriculture. In addition, the degree students, at the end of the third year of their four-year course, went on a week's agricultural study tour of parts of Europe. The staff took turns to lead this tour and I duly did my stint in 1988, but the tour was so successful and enjoyed so much by the students that I regretted I had not been able to travel with them more often. I also made it clear to the first-year students that they were expected to put on a student play towards the end of the first year. They always came up trumps and we had a play in each of my seven years in the Department. Naturally the standard varied, as one could not rely on each year having the same amount of dramatic talent. Some years the plays were very amateurish and slapstick (with the Head of School being the natural butt for the students merriment). When their years at Edinburgh were over, I was told on many

38. University Students 'Going Down Dinner', Edinburgh, June, 1988. Far right, Professor Ian Cunningham, guest speaker. Author, third from right.

occasions that these student plays had contributed a major part in binding the year together and that the friendships made in this way often lasted not only for the rest of the course, but also after leaving Edinburgh.

At the end of the final year the students, without any prompting, organised a 'Going-Down Dinner', to which my wife and I were always invited as guests. The standard of the speeches and the generosity of their praise for their lecturers and tutors were very re-assuring. However, Bunny and I always made a point of leaving well before the end of the proceedings, which invariably continued into the small hours. The students were invariably well turned out for the ceremonies at which they were awarded their respective degrees and diplomas. They were similarly well attired at the 'Going Down Dinners', and it never ceased to amaze me how students who dressed very casually for most of the year managed to look so immaculate on these two special occasions. The current Principal of Edinburgh University, Sir Stewart Sutherland, normally ends his address to the students at the University graduation ceremonies with the words: 'Now go out and celebrate; you have deserved it!' My agricultural students required no encouragement from me to celebrate; they got on with their merry-making unaided!

I have become an admirer of Scotland's national bard, Rabbie Burns. I
have already mentioned that, during my days with BOCM Silcock, I was
invited every January to attend the highly professional Burns Night Suppers
run by Jim Armitage, the Scottish Area General Manager, in one of the
Glasgow hotels. I decided, on my return to Edinburgh that I would
encourage the staff of the School of Agriculture to re-initiate Burns Suppers,
which had been organised a decade or two previously but which had fallen
into abeyance. I was assisted in this by a few very keen and knowledgeable
'Burns buffs' and together we devised a system whereby each Division in
turn took responsibility for a Burns Supper. Indeed, I encouraged a little
rivalry as to which Division could put on the best show. These Burns Nights
were a great success and I thought they had generated their own inbuilt
momentum, so I was a little disappointed to find that, as soon as I left, no
more Burns Suppers were organised. The students traditionally organised
Burns suppers but, as was to be expected, these were rather wild affairs with
the event being an excuse to over-indulge. I was regularly invited to these
students events and must have proposed the toast to *The Lassies* and *The
Druthie Cronies* many times. However, I always declined invitations to give

39. *Visit of Lord Provost of Edinburgh to the Scottish Centre for Agricultural
Engineering of the Edinburgh School of Agriculture, 1989. From left: The author,
The Lord Provost, Professor Brian Witney, Director, SCAE.*

The Immortal Memory, which I believe is a toast that should be taken seriously and not when it is difficult to make oneself heard above the din of a very merry company. Once again, the students were splendidly turned out for these Burns Suppers, with almost all the lads dressed in kilts and the lassies with long dresses and tartan sashes. I also arranged for a Burns Supper to be one of the important events during the celebration of the Centenary of the Chair of Agriculture in 1990. This event was held in the Edinburgh City Hall and was a very great and enjoyable occasion, well supported not only by the staff but also by many friends and well wishers attending as our official guests.

My final Burns Supper took place in the January following my retirement. It had been arranged months beforehand and was therefore already in the calendar. I was asked to make a special speech in the form of a new toast, *Farewell to the Haggis*, and I used the opportunity to thank my former colleagues in the School of Agriculture for all their support over the past seven years. Some of the verses, written to honour some of the folk mentioned in this book, went as follows:-

Laurence Beveridge

There was ae sang amang the rest
Aboon them a' it pleased me best
That to our Lawrence twas addrest
　　Plain truth to seek
His name excels for ever blest
　　No further seek.

Professor John Holmes

But ye whom social pleasure charms
Whose hearts the tide of kindness warms
Whose handicap is double ones
　　We must agree
To down a dram wi' all your charms
　　At nineteenth tee.

Mr Alec Barbour and Mr Sandy Inverarity

Ye chieftains a', ye knights an' squires
Conducting Boards like heav'nly choirs
An' doucely manage our affairs
　　We a' salute ye
You tak' awa our pain an' cares
　　For which we thank ye.

Financial cuts and the Annual Budget Exercise
My curse upon your fallin' budgets
Which mak's us a' come oot in fidgits
Your moneys fit for moanin' midgets
   Sad sicht tae see
The tricks o' knaves, or fash o' fools
   Ye must agree.

Mr Stuart Ainslie
But mark th' accountant, haggis-fed
The tremblin' earth resounds his tread
An money matters, it is said
   His quill may draw
The figures never coloured red
   Stuart's awa!

Conclusion
An' to conclude my silly rhyme
I'm scant o' verse an' scant o' time
To make a happy fire-side chime
   To weans an' wife
That's the true pathos an' sublime
   O' human life.

There were many contractual differences between the terms of service of colleagues on the University staff and those on the College establishment. Although I had to spend the greater part of my time dealing with the management of the larger College, I was nevertheless employed by the University and had a University contract which set out my retiral age at 65 years. My College colleagues also had the same age for retirement, which put them out of line with the civil service, who were the College paymasters. Civil servants retire at 60, and pressure was put on the Colleges to bring their staff retiral ages into line. This caused a great deal of difficulty and when 'voluntary early retirement' failed to solve the problem, we had to resort to compulsory redundancy. Not unnaturally, this was greatly resented and after we had gone through all the required complex procedures, some staff appealed against the ruling and took us to Industrial Tribunals, where some won their respective cases. However, although sums were awarded in compensation, we were not ordered to re-employ people over 60 and so the job we were required us to do was duly done. I was by then just over 64 years of age, and although there was no legal requirement for me to retire

40. Dinner at Borthwick Castle to mark the Bicentenary of the foundation of the Chair of Agriculture at Edinburgh University, 1990. Professor Colin Whittemore, in making a speech to mark the occasion, presents the author with a root crop in memory of the pioneering work done by the third Professor of Agriculture, John Wilson, in promoting the introduction of roots into Scotland.

early, I felt it would be wrong to continue to work until I was 65 since I had been the person ostensibly responsible for the whole difficult operation. I therefore retired early from my Chair at the University as from the end of that academic year.

The Government Agricultural Department seized on this opportunity finally to merge the three Scottish Colleges into one. My opposite number in Aberdeen very conveniently resigned to take up a new appointment as Principal of the Royal Agricultural College at Cirencester, whilst Ian Cunningham, my other opposite number in Auchincruive, had already taken normal retirement a year or so previously. That meant the way was clear to appoint a single Principal in charge of the unified Colleges and Phil Thomas, previously at the Hannah Research Institute in Ayr and then the Principal of the West College for a short time, was duly appointed.

I had not contemplated retiring early, so had made no plans for such an event. Thus on the day that I retired I had no other jobs in hand or in mind. However, this did not unduly worry me and I thought of all the many interesting things I had wished to do if only I had had the time to do them. Now, it seemed, time would be on my hands. But events overtook me yet again, and within a month of handing over the keys of my office and the official car to Lawrence Beveridge, I found that I was busier than ever. Retirement, I discovered, was not what it was supposed to be, but that is another story which I will unfold in the next chapter.

CHAPTER 13

So This is Retirement

'Where there is much desire to learn,
there of necessity will be much arguing, much writing, many opinions;
for opinion in good men is but knowledge in the making'

Milton

THERE ARE TWO WAYS to prepare for retirement. One, which I have observed is the case with former colleagues, is to spend the last few years of work actively planning for the event, putting out feelers for future employment and generally letting it be known that one is in the market for interesting and rewarding jobs. The second, which is the way I approached the situation, is to do absolutely nothing, with the result that, on the day one finally hands over the keys of the office, one hasn't the faintest idea what to do next. It naturally helps to be able to pass on one's responsibilities to a trusted friend and colleague, who, one is confident, will take over the helm and guide the ship through whatever stormy waters may lie ahead. In handing over to Lawrence Beveridge, I had complete faith that the College and the School would be in good and safe hands. Although we were the very best of close friends, I made it my business never to interfere with Lawrence's future freedom of movement. I therefore only returned to my old office when invited. Such planning as I had done was vague. I would, I thought, be able to start serious reading again; books (other than technical ones) had become a rare luxury, bought but not properly read. I also assumed that there would be time to take up two favourite pastimes, hill walking and ornithology. I had the maps and the bird books; all I needed, I thought, was time.

For the first week things proceeded in this delightfully unplanned manner. But at the end of the week I was invited to meet my colleagues at the Edinburgh Centre for Rural Research (ECRR) to discuss a 'proposition'. The ECRR was an informal body linking all the institutions (Colleges, Universities and Research Institutes) having a presence on the Bush Estate and pursuing a number of conjoint research projects. ECRR was the successor to a previous organisation, the Edinburgh Centre for Rural Economy (ECRE), which was run by a Committee and used to employ staff

189

under the supervision of a paid 'Secretary'. When I came to Edinburgh in 1984, I was told that one of my duties was to serve as ex-officio Vice Chairman of ECRE, the part-time Chairman being a retired Law Lord, Lord Robertson, who had held the position for 15 years.

The organisation was clumsy and expensive, costing the member organisations a total of about £400,000 a year, and it was clear that it had to be trimmed down. Most of this money was spent on estate maintenance, fence repairs and hedge- and grass-cutting, and only a small proportion for the promotion of conjoint research. Eventually the term of office of Lord Robertson came to an end and the University, which was responsible for the appointment, took back the 'estate duties' in house and revised the Constitution so as to bring it more into line with future needs. I was asked by the University Principal, John Burnett, to suggest the name of a successor to Lord Robertson, so I looked for someone with a business background. I knew that Mr Quintin Brown, the Managing Director of Scottish Agricultural Industries, was about to retire. I had known Quintin for several years and when I returned to Scotland, he had kindly invited me to several of his Company's functions and had been a minor benefactor of some of the developments I had introduced into the East of Scotland College of Agriculture. John Burnett invited Quintin and me to join him for a working lunch, as a result of which Quintin was invited to take on the role of Chairman, with me continuing to act as Vice Chairman. The new arrangement worked well and Quintin successfully introduced a programme of re-organisation. As part of this major change, the previous staff were paid off and the new body now operates mainly on a voluntary basis. It was necessary to adopt a new title to signify this radical change and the name was changed to 'The Edinburgh Centre for Rural Research' (ECRR). The Chairman since 1990 has been a senior academic in the Science Faculty of the University and is currently Dr Des Truman, a former Vice Dean of the Faculty of Science and at present an Assistant Principal of the University.

The new Board of ECRR resolved that the re-constituted body should have a part-time Scientific Director to take charge of the day-to-day operations and future planning. Would I take on the job? Having been the Vice Chairman for the previous six years, I knew exactly what was involved and I agreed to accept the offer. I did the job for six years and greatly enjoyed the work. It involved arranging a monthly meeting of the Directors, a different Institution in turn acting as host, in order to see what was going on and to have informal discussions about conjoint research programmes. In between, I invited VIPs to meet Directors at an informal lunch, which was preceded by a visit to selected research establishments of interest to the individual guest. These VIPs included Ministers of Agriculture, as well as important academics such as successive Principals of Edinburgh University and the Chief Executives of the Research Councils. After lunch, each Director in turn gave a snapshot of the work of his own Institution, indicating how it fitted into the network of the multi-disciplinary ECRR. The visitor was then invited to comment on what they had seen and these comments often led to lively discussions. I organised about half a dozen or so VIP visits a year.

The Directors held a formal AGM, at which I insisted that my appointment was reviewed. I also organised an Annual Dinner for the Directors, again at Roslin, when their wives were invited. I arranged for a well-known personality, such as the Minister of Agriculture to be the guest-of-honour and these dinners still continue. Happily, the Directors always invite back their retired colleagues to attend these get-togethers and to meet former friends and colleagues.

One of the Directors of the ECRR was Professor Ron Roberts, who was then the Head of the Institute of Aquaculture at Stirling University, and I shared his view that agriculture and aquaculture should co-operate more closely. Aquaculture, or fish farming, had a lot in common with intensive poultry farming, since there were many similarities in the fields of fish and poultry nutrition. I argued that there was much that one form of intensive livestock production could learn from the other, especially in connection with the increasing importance of environmental considerations. Ron clearly took these discussions to heart, for a little time later I received a letter from the Principal of Stirling University, John Forty, asking me to join the Advisory Committee of the Institute of Aquaculture. I accepted, and have been a member of this body ever since.

Ron Roberts took early retirement from the Institute and his place was taken by one of his former senior colleagues, Randolph Richards. The good work started by Ron has continued to flourish under Randolph's guidance, and I was asked to stay on as a member of the Advisory Committee. I must have been a bit of a thorn in the flesh of the Stirling University administration, as I was quite critical of the constraints which were then placed on the intricate finances of the Institute. With outlying fish farms and research units, and with a trading account in which fish stocks were being sold for profit as a means of augmenting the Institute's income, the accounts were very different from those of any other University Department at Stirling. The problems involved were, however, similar to the problems encountered by the Agriculture and Veterinary Departments' farming operations at Edinburgh. However, the University administrators cannot have thought too badly of me, for in 1999 I received a letter from the new University Principal, Andrew Miller, asking me if I would accept an Honorary Doctorate of Stirling University.

Another research establishment which I persuaded to join ECRR was the University Marine Biological Station at Millport (UMBSM), situated on the Isle of Cumbrae, off the Ayrshire coast. This site was even more difficult to visit than outlying laboratories, such as the Aquaculture Institute at Stirling, and I thought it would be impossible to get a reasonable turnout of ECRR Directors to such a remote spot. I came up with the idea of holding a meeting there on a Saturday and I invited Directors to bring their wives and families with them. This unusual formula worked like a charm and we mustered a good turnout of over twenty people. However, there were certain logistical difficulties to be overcome. There was a long wait for the ferry, as I had not expected a sunny Saturday to bring out so many visitors to the island. Some families wandered off to buy ice-creams and others to examine the goods on offer at a charity sale in a nearby church hall and consequently missed the Cumbrae ferry. They caught up with us later. During the lunch break, which was a picnic in the Station's grounds, some other families wandered off to explore the area and a search party had to be sent out to round them up. Family visits, I discovered, were somewhat more complicated to organise than normal lunch-time meetings of Directors.

One interesting example of 'serendipity' occurred during the course of this visit. When we were being shown round the laboratory specialising in the study of crustaceans, a member of staff involved in the demonstrations casually mentioned that the exo-skeleton of the crab and lobster was one of the most 'dirty' animal surfaces you could find. The shell was covered with every form of virus, bacteria and algal growth imaginable. But there was one exception: the crustacean's compound eye. This surface, we were told, was always clean and sterile. The veterinarians and civil engineers present pricked up their ears at this point. It would seem that there were natural secretions bathing the compound eyes of these creatures with some very powerful chemicals, which kept invasive organisms at bay. Many larger animals, such as horses, suffer greatly from eye infections and many glass surfaces in sophisticated instruments are notoriously difficult to keep clean of small organisms. If the secret of the crustacean's compound eye could be revealed, there might be useful spin-offs into both veterinary science and optical instrumentation. A number of leads were informally discussed as a result of this chance remark, some of which are still being followed up by relevant conjoint research. In retrospect, our visit to Cumbrae was voted a great success.

After seven years in the post of Scientific Director, I felt the time had come to move on and make way for somebody more recently involved with the running of one of ECRR's twenty or so member Institutions. At that time, Professor Ian Aitken was coming up to retirement from his post as Director of the Moredun Research Institute. The Moredun, as it is more generally known, was in the final stages of a major move from its previous location in the Gilmerton Road of Edinburgh to a 'green field' site on the ECRR land at Bush, and as a result Ian had made many close contacts with his opposite numbers of the other Bush-based Institutions and was an ideal person to take over the Scientific Director's job from me. We were both fortunate to have extremely good support from the two Honorary Secretaries who were responsible for the secretarial aspects of the business of the ECRR Board, firstly Murray Black and then, after Murray's tragic and untimely death, Lawrence Hodgson-Jones. The contributions which good and efficient secretaries make to the smooth running of organisations such as the ECRR are very important, and often go unacknowledged.

My University duties, as Head of the Agriculture Department, meant that I had to work very closely with the various Deans of the Faculty of Science, and I got to know them all very well. When I retired, the Dean was Professor John Mavor, Head of the Department of Electrical Engineering, and we had become close friends and colleagues. A week or so after I had accepted the ECRR post described above, John invited me to his office to discuss another 'proposition'. He was looking for a part-time Development Adviser whose duties would be similar to those involved in the ECRR post, facilitating better communications within the Faculty and arranging VIP visits. This seemed to me to be a good way to keep my contacts alive within the University, without in any way stepping on the toes of my successors. I accepted the job and had a very interesting four years in harness with John Mavor, until he left Edinburgh University to become Principal and Vice Chancellor of Napier University.

I had served the Institute of Biology as a Vice President during the period 1977-1979, when my chief job was to master-mind the complex legal negotiations to enable the Institute to obtain a Royal Charter. This involved numerous meetings with lawyers specialising in this work, which in turn meant dealing with the civil servants responsible for the administration of the proceedings of the Privy Council. I had assumed that this stint would conclude my voluntary work with this London-based Professional Society, but in 1991 the then President, Sir Colin Spedding, rang me up and enquired whether I would be willing to serve yet again as Vice President 'now that I was retired'. Colin was a close friend and colleague of mine and I knew he had great plans for the future development of the Institute and, in particular, extending its membership to more people in public and private research establishments and strengthening the work of its Local Branches. The idea of working in harness with Colin attracted me and I agreed. I was only one third of my way through my period of office when, in 1992, Colin told me he was having great difficulty in finding someone with the right experience to take on the more onerous job of Honorary Secretary. Would I agree to switching my job with the Institute to this more important post? This time I asked for time to think the matter through, but eventually and somewhat reluctantly, I agreed to take the job on. This involved a lot of London-based meetings – at least three a month – and also a lot of travel to the various Branches into which the Institute was divided, to meet their officers and sometimes to give a lecture on the work of the Institute. The job was quite exacting, but we did succeed in bringing about many important changes to the administrative structure of the Institute. In particular, we strengthened the co-ordination role of the Institute by bringing some two hundred more specialised bodies under the wing of the Biological Council, which the Institute serviced.

In 1988 I had been invited by the University Grants Committee (UGC) to become a member of their Agriculture Committee and to take part in the 'Visitations' which the Committee made to Universities offering Agriculture as a degree subject. I enjoyed this work, which was then relatively informal, and I visited many of the Agriculture Schools of the country during the four-year period of my office. The arrangement at that time was that we would visit an Institution 'by invitation' and spend about three days on the campus meeting staff and students and seeing the facilities. At the end of the visit, the Chairman would prepare a report, in consultation with his team. We would then meet the Dean and Vice Chancellor of the University and have an informal debriefing session, in which we discussed the important points such as the Institution's apparent strengths and weaknesses. These visits were appreciated and I believe helpful to the staff of the University concerned. Low cost and informality were the order of the day. Probably the failure of the system was that we were never invited to those institutions that most needed informed but impartial advice.

In 1992 the UGC was superseded by the Higher Education Funding Council (HEFC) and the whole exercise became much more formal. It was transformed into a major Research Assessment Exercise (RAE). I didn't fully appreciate this at the time and when invited to be Chairman of the Agriculture, Food and Veterinary Science panel of the RAE, I agreed, little knowing what I was taking on. This exercise took three years to prepare, execute and report. During that period my panel met fourteen times and I made about a dozen or so visits to Bristol or other convenient venues to meet the permanent HEFC officials and to have round-table meetings with my colleague-Chairmen of the other panels. A lot depended on the outcome. We had to assign five different grades to the work of each Institution evaluated. The relevant part of the HEFC funding, relating to the payment for research, depended on these grades. The Universities getting a grade 1 got little or nothing from the exercise, but those getting a grade 5 received a large amount of cash per member of 'research active staff', which could make a substantial difference to the University's income. The other grades got intermediate amounts, but the steps were not linear and favoured those at the upper end of the scale Not unnaturally those institutions who did well kept fairly quiet, but those which came out badly complained and sometimes appealed, and generally lobbied publicly about the alleged inequities of the system. Clearly the exercise wasn't perfect, but at least it issued a clear signal that Universities were in the business of the advancement of knowledge through research, and that those that were good at the job would benefit.

In 1996 the exercise was repeated and once again I was invited to chair the Agriculture, Food and Veterinary Science Panel. Having made various suggestions, through my panel, for the improvement of the exercise, I accepted. Once again I was caught up in an intense programme of meetings and dialogue with Bristol. This time the rules were more strict and the scope for interpretation of them by the panels was less, but thankfully the number of published papers we were required to read was reduced. However, the Universities now understood how much of their funding was dependent on the exercise. In consequence the lobbying and jockeying was stepped up and there was a lot of 'special pleading', both from subject groupings and from individual institutions. Not unnaturally, the more the HEFC tried to tighten up the rules, the more the back-room lawyers worked out ways of circumnavigating them for their own benefit. Some of the appeals depended more on arguing technicalities about the regulations than on the fairness of the final outcome. One trick that was tried was to 'buy in' eminent scientists from abroad by offering them sinecure posts, in return for which they gave a few lectures and brought with them four of their best research papers to 'add to the pile'. Only very few tried to beat the system in this way and my panel were united in holding their ground. I believe our gradings were generally agreed to be fair, although it was hard for those who came out poorly to accept that the exercise was objectively executed. One of the main complaints was that their own Institution 'was not represented on the panel', which completely missed the point that panel members were chosen for their experience and not to represent any Institution. In fact, when a panel member had any connection with any 'unit' being evaluated, he or she had to declare an interest and play no part in that part of the exercise. The final outcomes of the 1996 exercise were very well written up and HEFCE is to be congratulated upon the clarity of presentation of the detailed results (181).

Another RAE will take place in 2001. Having served for three terms on the appropriate committee, my time is now up and I was delighted to be able to hand over the reins of office to my Vice Chairman, Sir James Armour, who is now doing his best to comply with yet tighter regulations and constraints. I wish him well, but must admit to becoming more and more concerned at both the complexity and cost of the operation, which is now creaming off many millions of pounds from the cash available to the Funding Councils to support the UK University system. Like so many other attempts to make the Universities more accountable, the system is now reaching the stage where the costs seem to outweigh the benefits. Not only is this wasteful of precious money which could be better spent, but many thousands of staff hours are taken up with the form-filling and extensive bureaucracy of the operation.

I was elected a Fellow of the Royal Society of Edinburgh in 1987. This learned Society was instituted by Royal Charter in 1783, for *The Advancement of Learning and Useful Knowledge*. Unlike its sister body in London, founded some years earlier, it represents all branches of learning, the Arts and Humanities as well as the Sciences. It is a wholly independent body with charitable status, governed by a Council elected from among its 1200-strong Fellowship. In 1992 I was invited into the President's office and asked if I would agree to take one of the two parallel posts of 'Secretary to the Meetings'. This involved attending at least half of the Ordinary Meetings of the Society.

At the end of my three-year stint I was invited out to lunch and asked whether I would consider taking on the General Secretary's post in a year's time. This proposition was a very different kettle of fish. The General Secretary was one of the three senior officers of the Society, and was responsible for overseeing all the work of the Society once decisions had been made by Council. There was a full-time staff to do most of the actual day-to-day operations, but the General Secretary had to ensure that decisions were carried out and that the work of the Society was generally conducted in a timely and efficient manner. He was ex officio a member of almost all the Standing Committees of the Society and, in particular, a member of the Business Committee and the RSE Foundation (the charitable arm of the

Society) and was responsible for the 'pay and rations' of the twenty or so permanent staff. It was a great honour to be invited and I accepted. The headquarters of the RSE are in George Street, Edinburgh which was quite convenient to where we lived.

As seems to be my lot wherever I go, the RSE was at a stage of great change and major development. It had recently bought 'the place next door', no. 26 George Street (the old traditional offices were sited at nos. 22 and 24). The newly acquired space required not only major refurbishment, but also integration into the adjacent buildings by breaking through exterior walls and all this was going to cost a great deal of money. When I took the job on, the staff were reasonably confident that the cash would be forthcoming from the Lottery Fund, and a submission for a grant of about £1.4M had been made. For my first year we were regarding the future rather like Mr Micawber – hoping for something to turn up. When we were finally turned down we were in a real quandary. We possessed a large imposing building (a former bank headquarters) but we did not have the wherewithal to adapt it for our use. Suddenly, literally almost out of the blue, a generous anonymous donor came up with all the cash required and so we could proceed. For two years the place was in turmoil, as the old bank building was virtually gutted before being turned into a most excellent addition to our existing rooms next door. We organised a splendid Opening Ceremony, performed most graciously by HRH The Princess Royal, so we now have a suite of rooms and offices of which we and our successors can be truly proud.

But preoccupation with building operations was not the only matter that exercised our time and attention. The coming of the new Scottish Parliament, with its separate administration and a new crop of Ministers and Members, made us re-think our aim and purpose and 'revamp' our administration to match. We needed to make ourselves more relevant to Scottish needs and aspirations. This has led us to develop new fields of endeavour, such as initiating a more international outlook by evolving closer links with the National Academies of other countries with Scottish links. We took a leading role, in conjunction with Scottish Enterprise, to assist Scottish business to utilise modern technological research conducted by the Scottish Universities, and initiated two new linked activities which are called 'Technology Ventures' and 'Commercialisation'. We needed to respond to requests for expert help and opinions from a variety of bodies, both Scottish and UK, and we had to create a new small Department, with a full-time member of staff, to co-ordinate the responses by Fellows to these requests. We process about three such requests a month. Some of these requests are for advice on issues which are topical and controversial, such as the pros and cons of conducting research on GM foods. We have been engaging with a number of activities with the young people of Scotland, by organising such things as Mathematics Master classes for budding mathematicians, and lecture programmes for some of the remote secondary schools of Scotland. We have attracted more funds with which to provide Fellowships for post-doctoral workers and we have consolidated our links with other professional bodies and academies, especially the Royal Society of London, the British Academy and the Royal Irish Academy. So much change in so short a time has inevitably produced criticism from some quarters, but the vast majority of Fellows have been supportive of the Council's efforts to bring the Society successfully into the 21st century.

All this has been quite a difficult task, but there have been many helpful colleagues guiding the plough and a very able and conscientious staff to assist and service the Fellows on whom the ultimate responsibility for all this work must rest. The head of the permanent staff at RSE is Dr William Duncan, who seems to be capable of working a 25-hour day 8 days a week! William is ably assisted by his main lieutenant, Kate Ellis, who is in charge of the financial administration of the Society. This, during the period of the enlargement of the premises, has been very complex and at times extremely delicate. It has been an especial privilege working with the other Senior Officers, all of them outstanding high fliers in their own particular field. During my first three years I worked very closely indeed with the President, Professor Malcolm Jeeves and with the Treasurer, Sir Lewis Robertson. I like to think that the three of us formed a very good team, working in close harmony with the permanent staff. It was a great privilege to be in office at the same time as Malcolm and Lewis and I greatly enjoyed their friendship and many kindnesses.

Just how high the Fellows fly, or have flown, is sometimes only made known to me when I spend time editing the Obituary notices which are published in the Year Book, some thirty or so notices appearing each year. When one reads the full account of the lives of these former Fellows, one feels very humble, wishing one had been able to get to know them better when they were still alive.

At present we are giving much more attention to our public relations and endeavouring to make our work more widely known in Scotland. When we bought and refurbished no. 26 George Street, we acquired a rather magnificent dome which adorned the former bank. The dome is floodlit in the early evenings, and we erected some rather fine lettering, spelling out 'Royal Society of Edinburgh', just under the cupola. One of the older Fellows accosted me at a meeting of the Society. 'Why on earth do we want to put our name up in bright lights?' he asked. 'We have got along very well for over 200 years keeping ourselves to ourselves.' When I replied that we were trying to improve our corporate image and make our contribution to Scottish Society better known, he grunted and replied that some lights were much better kept firmly under bushels.

I can almost hear the reader muttering: 'All work and no play makes Jack a dull boy', but this description of my 'odd jobs' should not be misinterpreted. I have done many things in my retirement which I never had the time to do when I was paid to work a five- (or was it a seven- ?) day week. Bunny and I have followed two important principles since I handed over the keys of my office in 1990. The first is that, when the sun shines and neither of us have appointments, we disappear into the country to do some walking (with the help of those old maps) or some bird watching (with the aid of binoculars and a good bird book). When we have two or more days in a row free, we put some things in the car and take ourselves off to other parts of Scotland away from the telephone, the daily post and the Fax machine. Also at the beginning of each year, we score out at least five or six weeks in our diaries, for which we book holidays – usually, but not invariably, overseas. In addition to this I have taken up the ancient sport of curling, and I curl (not very well, but enthusiastically) for both the Edinburgh Rotary Curling Club and the Veterans Curling Club on the two different Edinburgh rinks. When the nights draw in and winter comes, Bunny and I play quite a bit of friendly bridge. So although I may not have prepared for retirement (and I certainly have earned very little extra money since I stopped working full time), I have enjoyed the last ten years immensely. We have made many new friends and done many new things. More than that, I have learnt a lot of 'new tricks', including how to work the word processor on which the manuscript of this book has been bashed out.

CHAPTER 14

Reflections on Farming

'Blessings on him who invented the food that appeases hunger'
Miguel de Cervantes

I N CHAPTER 3 I spelt out the reasons that led me to pursue an agricultural career. From a very early age, I had set my heart on devoting my working life to something connected with agriculture. Although prostitution may be regarded as the world's oldest profession, ever since man progressed from being a hunter/gatherer, his whole life has been dependent on farming. In many parts of the world where subsistence agriculture is still practised, man is both farmer and consumer and the inputs to such a basic system are relatively simple: clothing, tools, trinkets and basic commodities essential for survival, such as salt. In most parts of the world there is a dichotomy between those who produce food on the farm and those who consume it in the towns and villages. Until relatively recently, there was a common bond and a shared understanding of these two ways of life. Those living in towns visited the country regularly and probably had several relatives earning their living from the land. Townsfolk were supplied with food by farmers from the country, many of whom traded directly by bringing their food in person to markets in the urban areas, or even selling food direct to consumers by means of the milk round or butcher's cart. When I was small, I often saw the French onion seller on his bicycle calling at homes in our street with strings of onions on his back or in his basket.

The various rural crafts and trades were also known and understood by urban folk. The blacksmith had his forge in every town and village, and would shoe the horses both from the farm stable and from the city livery. Poulterers had an array of rabbits, pheasants, fat capons, ducks and geese hanging from a rail on the shop front. Most housewives, whether they lived in the town or the country, knew how to skin a rabbit and pluck and dress a chicken. In addition to this, there were certain symbols of the importance of agriculture well-rooted in city customs and events. The Chancellor of the House of Lords still takes his seat on the Woolsack, but few of the noble Lords sitting in front of him now realise that the Woolsack was chosen as a

200

token of the nation's wealth, which for several generations was represented symbolically in the form of wool.

I rather enjoy the anecdote of the rural Bishop who decided to go from parish to parish in his diocese to get to know the people under his charge. He travelled in informal dress, but took with him his crozier as a symbol of his episcopal office. Coming across a shepherd in his travels, he stopped by the farm gate to pass the time of day. 'You and I do the same sort of work,' he said. 'You tend your sheep, I look after my flock. I see you are taking bales of hay to your sheep – I am bringing my flock heavenly food.' Warming to his theme, he continued, 'And to show how similar our work is, here is my crozier, which is almost identical to your shepherd's crook!.' The shepherd thought that he should make some contribution to the conversation and ventured to ask the question: 'Aye, Bishop, that may be so. But tell me, how'd you get on at last tupping season?'

In my lifetime all this has changed. Trainloads of folk from the East End of London no longer go down to the hop fields of Kent each year to live in the villages and help bring in the hop harvest. Boys from the towns and villages no longer follow the reaper-and-binder or harvester in the corn fields to catch rabbits as they run from the uncut corn in the centre of the field. The fruit in the Vale of Evesham is now being increasingly harvested by machine rather than by coach-loads of helpers from the nearby towns. And whereas in my youth most milking was done by hand, with an average of about 15 cows per milker, almost all cows are now milked by machine, as a result of which the ratio is about one milker to 150 cows. Indeed, if modern trends continue, there will no longer be a requirement for milkers, as robot milking devices will enable cows to select the times when they will gratefully relieve themselves of the weight of milk in their full udders without a man in sight.

The average farm size has increased during my lifetime from about 50-100 acres to something nearer 500 acres and the labour force, instead of increasing with average farm size, has remained static or even decreased. George Self's small holding of about 30 acres, on which I worked during my school holidays, has long since disappeared, the land having been bought up to enable nearby farms to enlarge to a more economic size. All these changes have meant that the number of people with 'hands on' experience of farming is rapidly decreasing and knowledge of what makes farms tick is being concentrated in the 2 or 3 per cent of the population still engaged in agriculture. These trends started before the War, but the demand for more and more food during the difficult War years accelerated the process. Farm size increased and mechanisation took over from horse-drawn farming operations. True, there are almost as many equines in the UK today as there

41. Malnourished child suffering from chronic protein deficiency, Uganda, 1957.

were pre-War, but almost every heavy farm horse has now been replaced with a riding pony, ridden by a lad or lass from a town with no knowledge of farming apart from the brief glimpse of the day-to-day operations of the riding stables.

After the War, rationing continued for almost another decade and the post-War Governments of both political persuasions were firmly committed to a 'Food from Britain' policy. Indeed, as late as the 1960's, when I returned from 13 years in the tropics, the then Government produced an important White Paper on *Food from Our Own Resources*, in which a programme of grants and subsidies was spelt out aimed at increasing self-sufficiency and boosting crop yields. Farmers received encouragement to deliver the goods by means of the Annual Price Review, when Government fixed minimum prices for all important food crops, both plant and animal. When the market price fell for any reason below the agreed threshold, Government stepped in to bridge the gap. Farmers could and did plan with confidence and could plant cereal crops or breed farm animals in the full knowledge that when harvest time came round, the price for their produce was fully assured. But behind this confidence, made possible by a guaranteed market for the majority of farm produce, farmers also knew, and appreciated, that they were regarded by the townsfolk as the heroes. It was a noble thing to produce food, and even more noble to make two ears of corn grow where only one grew before. The removal of any hindrance to maximising farm output was regarded as a small price to pay for the certainty of a regular and assured food supply. Thus the sight of hedges being removed (with the aid of Government subsidy to increase field size), or of wet and acid lands being drained (with the aid of Government subsidy to bring marginal land into full production), or of old decrepit farm buildings being pulled down and rebuilt with an iron-frame supporting a corrugated iron roof (with the aid of a Government building grant) was generally quite acceptable. The country as a whole was proud to think that, whereas pre-War they were relatively dependent on imported food, in the 1960s and 1970s overall self-sufficiency had risen to 60 or 70 per cent. Farmers could walk tall and hold their heads high; the work they were doing was of national economic importance.

However, in the thirty years or so since those post-war days of agricultural prosperity, when it was difficult for a farmer to lose money and the very best agriculturists were making healthy profits, all this has changed. The prices for farm commodities are now fixed in Brussels and not in Whitehall and much of what the farmer receives depends on the strength, or weakness, of the pound sterling against the Euro. No longer are farmers paid for all they can produce but in some cases, such as milk, they are restricted to a quota, which, if exceeded, results in a heavy penalty. This vicious system has another nasty angle, in that the quotas can be bought and sold, and quite large sums of money are expended on buying these precious bits of paper, which are essential before extra litres can be lawfully marketed. There is no guarantee how long these quotas will be valid, and when eventually the system is ended, many farmers will end up owning worthless paper quotas.

Another strange and unnatural system handed down from Europe is 'Set aside'. This is a requirement upon farmers to leave a certain proportion of their land unfarmed – indeed, to allow it to revert to bush and scrub. For agriculturists who have been brought up to husband their land well, the very thought of leaving it to revert to a cover of non-productive plants is alien to their character. How can it be, they ask, that a few years ago the Government exhorted them to maximise production, but now they are asked to allow their land to lie uncultivated? When 'Set aside' was first introduced, farmers were literally ashamed of the state of their purposely neglected land, which soon became overgrown with pernicious weeds. Later the rules were

42. Removing hedgerows with the aid of Government Grant, 1941.

changed so that certain scheduled weeds, such as ragwort and creeping thistle, could be controlled once official permission had been obtained.

This was not the end of the farmers' misery and woe. Next on the scene were the 'Animal rights' lobby, who claimed that all systems of intensive or semi-intensive animal husbandry were morally wrong, as they disregarded the precepts of animal welfare. First under attack were the intensive poultry producers, those keeping laying hens for egg production and others growing broiler chickens for poultry meat. The laying battery and the broiler house had been advocated and encouraged by government advisers employed by ADAS and the Scottish Colleges. Some of the infra-structures were state-subsidised and economic advisers were praising the virtues of large-scale production and consequent greater efficiency. Having taken this 'official' advice and discarded their old, more environmentally friendly, methods, long-established farmers were now being unfairly blamed as the instigators of unacceptable 'factory farming' methods of production.

After the initial attack on intensive poultry production, the 'Welfare Lobby' turned their attention to pigs. Pigs are often kept in stalls to prevent injury from fighting and ensure they receive their correct ration of food. Farrowing crates are used to confine a sow just before and after farrowing to prevent the sow from crushing some of her piglets as she lies down. In the traditional system of keeping sows it was regarded as normal for one or two piglets in every litter to be killed or injured in this way and the farrowing crate was devised as a major step forward by reducing this piglet mortality. Put another way, the farrowing crate was 'welfare friendly'.

An interesting incident occurred at the Dorset Institute of Agriculture in 1980, when the animal welfare protagonists were launching their crusade against intensive pig production. I was speaking at the annual prize-giving of the Institute and the ceremony was preceded by a luncheon at which I found myself sitting next to the Principal's wife. 'Did you hear the disturbance this morning when the Bristol University students barricaded the front gates?' she asked me. When I told her I didn't know there had been a student demonstration, she explained to me. They came here to protest about our intensive pigs, but when I approached the ring-leader to find out what was going on, I asked her to be specific about what the demonstrators found objectionable. She was told they didn't know, but that it was common knowledge that pigs were kept under barbaric conditions and that they had come to demonstrate about it. 'Would you like to see our pigs?' they were asked, 'then you can tell me what you find barbaric.' The ring-leaders agreed, and the Principal's wife led the way to the piggery and showed them the pigs. It was the first time any of the students had been inside a piggery and they were impressed, especially when the reasons for everything were explained and it was pointed out that, in the winter period, it was far more comfortable and healthy for pigs to be confined indoors than running around in the mud on a cold wet day. Duly chastened, the student leaders called off the demonstration. 'What subject are you studying at University?' the Principal's wife enquired. 'Sociology', replied the ring-leader. 'How interesting', said the Principal's wife, 'then I think your energies would be far better spent campaigning against cruelty to children instead of something you clearly know nothing about!' 'Bully for you', I thought, when she recounted this incident.

We are said to be a nation of animal lovers. The membership, and income, of both the RSPCA and the RSPB far outstrips that of the NSPCC (which does not even rate the title 'Royal'). Now both these former bodies do excellent work and I support them in their important activities, but there are many other more extremist organisations in the so-called 'Animal Rights' movement whose actions are not only illegal but run contrary to the cause they allegedly support. The freeing of mink from mink farms to run amok in the countryside; the release of chickens so that they are mown down by vehicles on public roads; the mindless torching of research institutes working for the well-being of both animals and man. This does not make any kind of sense and seems to bring out the very worst features of human behaviour rather than showing true concern for animal welfare. I personally have never liked laying batteries, certainly when three or more birds are kept in a single small cage. I personally do not eat veal as I dislike the system of rearing calves on an iron-free diet merely in order to get white instead of pink meat. I have always preferred straw yards to sow stalls, but I realise only too well that when strange sows are mixed together in an open yard, horrific injuries can result from fighting. My point is that the motivation of the farmer, when introducing these new and efficient systems to his farm, has either been as a response to the call by previous Governments to increase output and efficiency for the good of the nation, or else have been introduced for the good of the animals under his care.

There seems no end to the stresses and strains imposed on the farmer today. The next attack on his farming methods has come from the environmentalists. Many of us were sorry to see hedgerows disappear in East Anglia and the wetlands drained in the Somerset levels. Once again, however, there has been little understanding of the fact that these practices were a direct response to Government pressures, official advice and financial incentives in the War and post-War years. The layman believes, not unnaturally, that these changes in animal housing have come about solely at the wish of the avaricious farmer.

Finally (or is it really finally?) on top of all this came BSE. The layman, unfamiliar with either ovine spongiform encephalopy (scrapie) in sheep, or bovine spongiform encephalopy ('Mad Cow Disease') in cattle, has been led by the media to think that one or both of these conditions have been brought about by the unnatural feeding of 'parts of dead animals' to livestock by the wicked farmer. It is not generally known that scrapie has been around for at least two hundred years. Neither is it generally known that meat and bone meal has been fed to farm animals for three quarters of a century. We also happily ignore the fact that we have been eating meat and offal ('parts of dead animals') since time immemorial, and that black pudding and haggis are considered common delicacies, particularly in Scotland.

'But', the public naturally wish to know, 'how is that BSE has only been found since about 1987, and is not BSE exactly the same thing as Creutzfeldt-Jakob Disease (CJD) in humans? Are they not both connected with the feeding of meat-and-bone-meal to cattle?' It is difficult to provide clear-cut answers as to the cause and spread of BSE. However, the following comments may be of interest. They are not generally known and as a consequence the farmer is blamed for something outside his control.

For decades meat-and-bone meal has been made in specialised plants. There is a pan-European trade in the commodity, which is imported and exported according to supply and demand. In the past, the chief hazard with this material was the contamination with coliform bacteria and other organisms which can cause diarrhoea and other disorders in animals and man (e.g. *Salmonella* and *E.coli*). These disease organisms were destroyed by the traditional high-temperature process by which meat-and-bone-meal was manufactured. The technique relied on solvent-extraction to remove the fat from the product, which could then be used for other purposes such as the manufacture of soaps and detergents. This original process would have effectively killed the agents now thought to be responsible for BSE – prions. These prions are a sort of primitive protein, much smaller than normal proteins. They are not viruses or 'living organisms' although they appear to behave rather like viruses and seem to be capable of reproducing themselves when the conditions are suitable.

Unfortunately, the solvent extraction process is 'dangerous' from the point of view of human safety, since solvents are volatile and highly inflammable and many explosions occurred in meat and bone meal plants, leading to loss of life or serious injury. For this reason, the Health and Safety Executive brought pressure to bear on the plant owners to alter the procedure to a heat treatment process, without the use of the dangerous solvent. This method was a 'Low temperature – long term' heat process, capable of killing off all the then known dangerous pathogens, but which did not unduly damage the proteins. These were required to remain fairly intact if they were to be a nutritious protein feed for animals. At the time the changes were made (in the late 1970s) BSE (and prions) were unknown and so it was impossible with the knowledge then available to check whether or not the process would destroy prions. It transpired that it did not and apparently, unfortunately for all concerned, these prions can pass intact through the new process and are now thought to have been the cause of the outbreak of BSE in cattle. It is further believed that the prions were derived from parts of the nervous system of sheep containing the agents of 'Scrapie'. Since BSE in cattle has a long incubation period, it was not until the late 1980's that it was first recognised and recorded.

Extremely good detective work carried out in veterinary laboratories then pointed the finger at a common factor: the consumption by BSE-infected cattle of meat-and-bone-meal at an early stage in their lives. Once this was known, meat-and-bone-meal was withdrawn by law from cattle feeds, but not from feeds destined for other livestock. Whereas the law could be enforced in factories, which were regularly inspected by government officials, it was impossible to ensure that meat-and-bone meal was not still being used by farmers making up their own rations. Meat-and-bone-meal is only used at low levels of inclusion, usually less than 5 per cent, so if a farmer had purchased several tons of meat-and-bone-meal it could take a long time before the last bags of the material were mixed into animal feeds. Although farmers were told not to use the material for cattle rations, it was still permissible to use it for other stock.

Other questions arose which were even more difficult to answer. Could BSE be passed from a cow to her calf? Could BSE be transmitted to sheep (from which it apparently came in the first place) and then relayed back to cattle, perhaps grazing in the same field? Could a cow with BSE infect other cows with which she came in contact? These questions needed long-term experimentation, which is only now coming to fruition and providing answers.

In 1989, soon after BSE had been first recorded, I was asked by the Milk Marketing Board (MMB) to chair a committee looking into the matter of cattle identification and recording. Up to then, responsibility for tagging cattle and keeping accurate breeding records was partly the responsibility of the MMB for farms using their Milk Recording Service (National Milk Records) and partly the concern of the animal breeding societies, who required this information for their pedigree registers. There were many cattle who did not get recorded by either party, as the farmers concerned neither kept pedigree stock nor bothered to record individual cow milk-yields.

I assembled an excellent team, all of whom were prepared to work long and hard in order to produce a Report within the six-month period allotted to the task. Our diligent Secretary was Frank Armitage, 'on loan' to us from the Milk Marketing Board. The other members were: Chris Bourchier, a senior member of ADAS, David Leaver (Professor of Agriculture at Wye College), Tom Kelly (from National Milk Records) and Duncan Spring (Chief Executive of the Holstein-Friesian Cattle Society). Chris Bourchier and David Leaver are now both Trustees of the Frank Parkinson Agricultural Trust (see chapter 16).

We were set up in May 1990 and completed our Report in December, 1990 (170). We held over 20 meetings, received 106 written submissions and interviewed 48 of those providing written evidence. When the report was published, we attended meetings with eight of the main interested parties, to deal with questions and ease the acceptability of our recommendations. Not one interested party dissented from our findings and we prided ourselves that nothing stood in the way of implementation. We reckoned without the Treasury. Of our twelve recommendations, the last one read, 'That the Ministry of Agriculture, Fisheries and Food (MAFF) be encouraged to promote the formation of the Cattle Data Centre (CDC) and to assist in obtaining funds to set up the CDC and its constituent database for the public good'. The 'price tag' on this recommendation was £5m. We were at pains to emphasise that this was a 'one off' cost, and recommendation 3 stated, 'That the CDC be self-financing by 1995'.

In the event the Treasury refused to finance part of the cost of the computerised data base of the CDC . They apparently were not given clear-cut objective advice from MAFF, since there appeared to be a difference of opinion between the animal scientists (who supported our report) and the vets (who did not). One of our arguments in support of the CDC was that the breeding and movement data would enable the government to carry out detective work on any future new condition affecting cattle disease more expeditiously. The vets argued that the 'worst case scenario' for a major disease outbreak was Foot and Mouth Disease and as the control of this condition did not need a Central Cattle Data Base, they could not envisage any scenario that would. The Treasurer took the least-cost advice from MAFF and thereby saved the taxpayer about £5m, but over the period of the next ten years they cost that same taxpayer a figure of between £5b and £10b, because of the non-traceability of BSE cattle over that period. It is, of course, easy to be wise after the event, but it is sad to think that these difficult decisions are too often made on the basis of apparent costs rather than on an objective and holistic appraisal of the problem.

So, whither agriculture? When I was a schoolboy during the War, the farmers were the nation's heroes. When I was at University in the post-War days, agriculture was regarded as a worthy occupation, which not only brought reasonable financial rewards but which gave great job-satisfaction. When I returned from abroad, farming was having to tighten its belt and become more efficient, but there were still fair rewards for the top 75 per cent of farmers and the public were still on their side. Over the last two decades I have seen farmers relegated from the ranks of the 'good guys' to those of the 'baddies'. Many farmers have gone bust and others have committed suicide. A recent TV programme, *The Dying Breed*, reported these recent chapters in farming history very well. My heart truly grieves for an industry parts of which have died, whilst other parts are slowly dying, as a direct result of public apathy, government indifference and increasing harassment by the ill-informed anti-farming lobbies of various shapes and sizes.

The outlook for many farmers is bleak indeed, but agriculture is an evolving art aided by a rapidly developing science and whilst some aspects are withering away, others are producing new shoots. Change is the order of the day and those farmers who are willing and able to contemplate change and move into fresh areas and adopt new practices will survive and prosper. The future industry will require fewer farmers and even less agricultural workers and farm size will increase correspondingly. Fortunately for those who remain in the business, the world population will always require an ample supply of good food. There is an old Chinese proverb which translates, *Those who have food have many problems; those that have no food have only one.* Two of the most basic necessities in life are food and water. Miguel de Cervantes was quite right when he wrote, *Blessings on him who invented the food that appeases hunger.*

CHAPTER 15

Whither Education?

'We live and learn, but not the wiser grow'
John Pomfret

DURING MY TEACHING CAREER I have witnessed many changes in the structure and content of education. I have been involved personally in many of these changes. In others I have noted developments at one step removed, either by observing events as our children and grandchildren went through school to Higher Education, or by having friends who are members of the teaching profession. I will deal with these changes in terms of School, Technical School, Polytechnic and University.

School

In Chapter 2 I described my own War-time school days at Whitgift School, where the senior boys, especially the prefects, were responsible for keeping discipline not only in the corridors and playgrounds, but even in the nearby town as well. Students caught in Croydon without the obligatory school cap, or with collar and tie undone, would be duly summoned to attend a meeting of the Prefects' Court. The staff, on the other hand, were responsible for discipline inside the classroom, and naturally the Headmaster and his assistant had an oversight of discipline for the school as a whole. In the 1940s, many schools, both in the public and private sector, were run on similar lines. This has all changed, particularly in the public sector. Schoolchildren are now judged too immature to take on these responsibilities and the result is that a much heavier burden devolves onto the staff. In the public sector, I suspect that the Teachers' Unions would oppose the use of 'student labour', as they would view this as a means of cutting down staff numbers.

In recent years the matter has become more complex, especially in the case of schools where disruptive children are a major problem. Corporal punishment is out. Whereas a teacher can no longer exercise any form of physical control over unruly children, the pupils sometimes abuse their new freedom from corporal punishment by verbally or physically attacking the staff. Now I fully understand that corporal punishment, or indeed any form of physical correction, is a tactic that should only be employed as a last

*43. Headmaster and Senior Prefects of Whitgift School, 1946. The author is fourth
from the right in the second row.*

resort. I also appreciate that the skill of effective teaching is to make the
lessons interesting and relevant to the children, so that they are not
perceived as dull and divorced from any connection with normal daily life. I
am full of admiration for the few gifted teachers who manage to control a
class of potentially unruly and disruptive children by force of their
personality. However, I am concerned at the situation, which appears to be
becoming increasingly prevalent, in which the class is one long, unhappy
battle between teachers close to losing their authority, and a minority group
of students whose one objective is to provoke as much disruption as
possible, so that their classmates who wish to learn are effectively prevented
from so doing.

The old adage 'save the rod and spoil the child' is now politically incorrect
and indeed its implementation would result in a legal action for assault
against the teacher concerned. However, society does not appear to have
produced an answer to the teachers' oft-repeated question, 'If the class do not
behave as normal human beings, what action can I take?' Even the ultimate
sanction in my time – expulsion from school – is no longer a satisfactory
option, since exclusion can result in problem pupils being moved from
school to school. Often the children themselves 'vote with their feet' and
absent themselves from school, knowing that there is little change of being
caught. Indeed I suspect that there are some teachers who are not sorry to

44. Girls dining in the refectory at St.Andrew's School, Bedford, 1967.

see the difficult children stay away from school, rather than break-up the good order of an otherwise well-behaved class.

Another change since my school days is in the nature and conduct of public examinations. In the War years, at 16 years of age, most English secondary school students took the 'School Certificate' examination and, a couple of years later, if they stayed on at school, the 'Higher Schools Certificate', in a range of subjects according to student choice and academic ability. The 'Matriculation' qualification, which was the entrance examination to University pre-war, necessitated passing six subjects, all at the same time, in the School Certificate examination. English, Mathematics and Latin and one Science subject were obligatory, as also were either History or Geography and one Modern Language. This was a tall order and not all students managed to pass all six subjects. Furthermore, most Universities demanded more than the minimum of Matriculation and required various subjects to be taken at the more advanced level of the Higher School Certificate. If one passed enough subjects well enough in this more advanced examination, one could usually obtain exemption from the first year of a University course at London and other 'red brick' Universities, although at Oxford and Cambridge the full course in residence was obligatory. Doubtless the system was unfair on those children who were 'late starters', but it was an effective way of sorting out those who, at that time, might benefit from Higher

Education and those who could not. Maybe insufficient allowances were made for such circumstances as sickness (or perhaps one's home being destroyed in an air raid), but all in all, the system did not do too bad a job. Since then, the public examination system has changed at least four times. English Matriculation is dead and the lower standard pupils can now acquire the necessary pieces of paper by taking the subjects one at a time, repeating those that they fail, and sometimes not taking any extensive examination as such, but moving up the ladder by 'continuous assessment'.

The result of all this is that some school children now arrive at University without ever having taken a 'total test' equivalent to Matriculation. Moreover, since the ones with less academic ability would find it very difficult to pass an old-style rigorous degree examination, these 'diets' have followed down the same path. 'Multiple Choice Questions', 'Continuous Assessments' and permission to bring text books and calculating machines into the examination room have resulted in a very different University Degree Examination. It would seem to me that many of these changes have been dictated by events that started at school and have, for convenience rather than by design, been carried forward onto the University Campus.

Technical School

As mentioned earlier in this book, my first 'hands on' teaching experience was at the Canterbury Technical School. This was a 'boys only' school, and the students placed under my charge were deemed only fit to 'do' agriculture. The system clearly failed to give them a good training in subjects such as agriculture, which were deemed to be 'non-academic'. Although, through hard experience, I acquired the necessary skills for keeping a class 'gainfully occupied', I cannot claim I made any impact on the boys' agricultural education. The system might have been perfectly satisfactory for potential plumbers or electricians, but it certainly did not help future agricultural workers, who would have been far better catered for at a later stage in a county agricultural college.

'Polytechnics'

On my return to the UK from abroad, I found that well-equipped Polytechnics had arrived on the scene and were thriving. They managed to attract excellent staff; they were given good 'technical' teaching facilities both in terms of buildings and equipment and although inevitably they always asked for more, they were secure in the knowledge that they were doing a good job which was important to the nation and appreciated by their students. My work with BOCM Silcock brought me into contact with many institutions of this type and I developed a very healthy respect for them. Their students were well motivated and left their 'Polytechnic or Monotechnic College' very well equipped to enter their chosen profession. The Polytechnics usually included a 12-month placement in industry, whereby a good student was often assured of a job at the end of his course, as employers were carefully watching their 'placement students' for future potential as employees. The Polytechnic staff undertook some experimental work in their spare time, but this was by way of an 'optional extra' and not one of their key tasks. The experimental work was usually more development-orientated than basic research and was often done in response to requests by the companies to which they were sending their students for the placement year. There was therefore an excellent rapport between the staff of these institutions and the commercial world. The Polytechnics were judged by the manner in which their students satisfied their employers. There were no complicated external audits or controls superimposed on the system. The 'industry' they served knew exactly where to go for good-quality students and the feed-back, although informal, was effective.

In the 1980s, when I was back in Edinburgh, one of the first jobs given to me by my two COSAC colleagues was to represent them at meetings of the Council of Principals and Directors of the Central Institutions (COPADOCI) of Scotland. Polytechnics, at that time, were known in Scotland as 'Central Institutions' or CIs, which meant they could recruit students from all over Scotland – not just from their local area. They were regulated by the Education Department of the Scottish Office, which differentiated them from local County Colleges, these latter being managed by the Local Authority. The SAC was a member of COPADOCI by virtue of the fact that it was regarded as a 'CI', although it had no connection whatever with the Scottish Education Department. Our masters were the Department of Agriculture and Fisheries for Scotland (DAFS) and as such I was 'the odd man out'. In addition, my own contract of appointment was with the University and not with the Scottish Office, so there were two major points of difference between me and my twenty or so COPADOCI colleagues.

I was impressed by the dedication of the CI Principals to the task of Higher Education. They were responsible for a diverse collection of disciplines and managed large and well-trained staffs. In many ways, I envied them their single-minded purpose in life, which was teaching to a high standard students who, for one reason or another, had opted for a Polytechnic type of Higher Education. On the other hand, my SAC colleagues and I had to juggle the three main tasks of teaching, research and advisory work. This sometimes made for divisions within the staff and there were occasional unhelpful arguments as to which of the three responsibilities was most important. Although there were many merits in the SAC system, there were also complications which did not bedevil my Polytechnic friends on COPADOCI.

However, in the mid-1980s tensions developed within the CI system. The Principals were unhappy at their perceived lower standing *vis a vis* the Scottish Universities. They argued that they would lose out in the battle for student numbers if they did not possess the magical accolade, 'full University status'. Meetings with the Scottish Office officials were often dominated by long speeches in favour of breaking down the barriers of the binary system and of making all Polytechnics full Universities in their own right. I was always sceptical as to whether this move would be in the best long-term interest of the Polytechnics. As CIs, they had a clear-cut role to play in Higher Education. They positioned themselves near to Industry and Commerce and they produced students ready and able to take part in many of the trades and professions important to the creation of the wealth of Scotland. Being 'only' a Polytechnic did not prevent them from doing this, and doing it well. But it was clearly a matter of pride as well as of principle and the lobbying went on, inside and outside COPADOCI, until eventually the political will was there to make the necessary transformation. Almost overnight, a whole new crop of Universities was born.

But while these moves were gathering apace, the Universities were also 'reconsidering their position' – or, rather the politicians were re-considering it for them! Universities, it was argued, must be at the forefront of knowledge of the subjects they encompass, and this means producing research outputs up to national and international standards. The 'Research Assessment Exercise' (or RAE as it became known) was born. I was party to the working of this exercise, and I describe this experience in chapter 16. This operation had one prime purpose: to evaluate research excellence at all UK Universities and to reward the ones deemed to be producing the best Research and Development. The Polytechnics, having blindly followed the University lead, were naturally dragged into this exercise. Unsurprisingly, their RAE ratings were initially very disappointing. Their staffs were caught on the wrong foot. They had been attracted into the Polytechnic system because of their skills in teaching and not many of them had a track record in top-quality academic research. However, 'Needs must when the devil drives' and their Principals applied pressure to step up the quantity and quality of staff research. This meant some good staff failing to meet the new job specification and other staff being recruited, not so much for their excellence in teaching but primarily for their past research record. Many University research workers at the Senior Lectureship/Reader level were attracted to the new system with an immediate offer of a Professorship. They brought with them, as expected, their research skills, but they were not necessarily the skilled teachers that Polytechnics had recruited in the past. In general the standard of teaching at some of the former Polytechnics tended to drift down whilst the standard of research rose. Perhaps this shift was a necessary part of the package of entering the ranks of the University system. Whether it was desirable for the future good of the students concerned is another matter.

University

When I was a student at Wye College, I was aware that the finances were dictated by a major exercise known as the 'Quinquennial Review'. Once completed, the results of this study guaranteed the funding for the next five years, with various allowances added on for inflation. I can recall both the Principal and the College Secretary complaining that five years was far too short a period to enable decent long-term planning. With the present system, the maximum period of stability is normally just under a year. The root of the many problems facing Universities today is still funding, but instead of a stable five-year period allowing for good medium-term planning for normal growth and expansion, the time horizon has been drastically reduced. However, the expectation is that there will be no more money for expansion but instead a gradually diminishing income, since the Government wish the Universities to make regular 'efficiency gains' – a clear implication that in the past they had been working inefficiently, if not improperly.

It is difficult if not impossible to judge the efficiency of a University. The accountants would have us look at 'cost per student place' or 'percentage of students completing their courses'. Such parameters are badly flawed, as are all others which attempt to over-simplify this complex area. Courses differ in costs simply because different subjects need radically different facilities. The most expensive University course to run is a Veterinary course. This is simply because all students must gain experience of a wide range of animals, alive and dead, unwell and healthy. Medical courses face the same problem, but here many costs can be offset against the provision of high-level consultancy facilities for the NHS. There is no NHS for animals! The lowest-cost courses could well be in subjects such as modern languages, although even here, with 'language laboratories' and complicated technologies for unravelling phonetics and analysing different dialects, the costs are rising fast. It clearly does not follow that Language Departments are more efficient because they are less expensive, nor does it follow that they are getting inefficient when their equipment costs rise because the methods are becoming more sophisticated.

When it comes to measuring 'outputs' in terms such as student success rates, the matter is even more difficult. It is relatively easy to boost student pass rates by lowering standards, but does this increase real efficiency, whatever that term may mean? Most would argue (and I am one) that standards must be retained at all costs. But if undue criticism is made about inadequate student numbers and if the response to this is to accept more students from 'unusual' scholastic backgrounds, or with lower grades in school examinations, then it is inevitable that pass rates will fall if standards are maintained. The Universities are in a cleft stick, especially the 'new' Universities who, a decade ago, were Polytechnics. Holding or raising standards inevitably means more funding difficulties. The way out of funding difficulties is to take in more students and lower standards. The older Universities (and a few of the newer ones) can improve their situation by massive increases in research funding, and by majoring on courses where student demand is buoyant. In the future the newer Universities could be in real trouble, primarily as a result of financial problems, and it is likely that there will be several more mergers or even 'take-overs' in the next decade or so.

This chapter asked the question, 'Whither education?' I do not know the answer, nor, I suspect, does anyone. But three things are clear. First, the political will to allow a higher proportion of the population to enter University may be a desirable aim, but it has brought major stresses to bear on those who deliver the mass higher education now required. Second, it has not yet been demonstrated that the extra numbers of students now passing through Higher Education are opting for subjects in the right proportions necessary for the public good. It is clear that subjects perceived by pupils to be 'difficult' and therefore less popular, like engineering, are being under-subscribed, whilst more glamorous subjects, such as sports management and media studies, are in danger of being over-subscribed. Lastly, the rush to create numerous new Universities may not have been so well thought through as it might have been. It may well come to pass in the future that a new sort of 'Binary Divide' will appear. On the one side will be those Universities at the forefront of research and scholarship endeavour, with world-wide reputations for excellence. On the other will be Institutions which, although awarding degrees, are expending unnecessary effort in order to compete in terms of academic excellence with the traditional Universities, with the consequence that student teaching and welfare will inevitably suffer.

Nevertheless, changes will still come about and the educational institutions must and will respond to them. But, perhaps those responsible in high places for the policy of Higher Education should heed the words that John Pomfret wrote as long ago as the seventeenth century, '*We live and learn, but not the wiser grow*'.

CHAPTER 16

Councils and Committees

Oh, give me your pity
I'm on a Committee.

Anon.

MY INTRODUCTION TO committee work started at an early age. When I was about 12, Arthur Stevens, the Curate of my church at Sanderstead, started a Youth Fellowship and I was elected its first Secretary. I remained in that post for five years, up to the time I went to University, and my secretarial duties taught me a great deal about the pros and cons of committee work. I enjoyed the responsibility and learnt a great deal the 'hard way', especially the important lesson that one can get a lot more constructive effort from a committee if one spends ample time, behind the scenes, preparing the way. I suppose in political terms this would be called lobbying, but it was not done with any pre-arranged solution in mind, but merely to make sure that decisions were made without too much procrastination and loss of time.

In parallel with this youth work, I was on a few committees at Whitgift School. As already mentioned in chapter 2, I was Chairman of the School Book Selection Committee and soon discovered the importance of preparing the work before taking a meeting. The boys took the privilege of selecting books for the school library very conscientiously. Woe betide the Chairman if he had not read at least all the reviews, if not some of the actual books, which were on the short list for purchasing. I had also been a Senior Prefect at the School and as such a member of the Prefects' Court. I was impressed with the time taken to consider each boy's misdemeanours and ensure, as far as possible, that the punishment fitted the crime

At Wye College we seemed to run most activities without a heavy overlay of committee work and the two committees that I recall sitting on were both external to the College. One was the Management Committee of the village Youth Club, which I started with a college friend, Eric Cordell. This Committee was chaired by the Vicar of Wye, but as Youth Club Leaders, a lot of the administrative work of running the Club fell on our shoulders. A more exacting committee job was being on the Council of the Southern Region of the Youth Hostels Association (YHA) of England and Wales. I was on this body by virtue of the fact that I was the Leader of the Wye College YHA

220

Group. The Council met in London once a month, which meant that I got my return rail fare paid for by the YHA and so was able to spend some time in London at someone else's expense. The Council was responsible for the care and maintenance of the hostels in its Region and we were expected to get to know as many of the establishments as we could manage to visit. The YHA started a hostel on a farm at High Halden and, being an agricultural student, I was expected to contribute some useful inputs into the efficient running of the farm. Unfortunately, the Warden of the hostel, who was not all that successful in balancing his two jobs of hostel management and farming, did not take too kindly to what he regarded as 'outside interference' and I don't think I made the slightest impact on his farm policy or his profits.

My year at Edinburgh was, as far as I can remember, entirely free of committee work. I was obviously much more concerned with settling into married life and doing well in my postgraduate Genetics course than in finding time for other commitments. This was probably the most 'committee-free' stage of my career, although I did not realise it at the time.

In Uganda my wife and I became joint Secretaries of the Serere Staff Club and were responsible for bulk ordering of groceries on behalf of all the staff and for stocking and running the Club House bar. This was hard work and had its difficult moments. On one occasion we were trying to balance the Club books and found we were somehow 10cts out. The temptation was simply to put 10cts. into the cash takings but we were determined to find the source of the error and spent a very long evening wrestling with the accounts, followed by a sleepless night worrying about them. We never found the missing 10cts! We also had to pacify our customers when they complained. Even dried tinned goods did not keep forever in the tropics. An angry house-wife whose bread did not rise because the dried yeast had gone off was a fearsome adversary.

A couple of years after my arrival in Uganda, I was put onto the Development Committee of the Government Agricultural Department, responsible for overseeing the animal parts of the country's R&D programme. This gave me an opportunity to get round to many parts of Uganda I had not previously visited. It was interesting work and my chief contribution was to assist in working out a cattle breeding programme for Uganda. Such programmes are essentially long-term, so it was only really beginning to produce results when we left the country in 1957. Part of the programme entailed importing some high-grade Kenena cattle from the Sudan, which were in danger of being slaughtered for meat during the civil war then raging in the south of that troubled country. We successfully brought some surviving members of this breed over the border into Uganda and located them at the Government Stock Farm at Entebbe. I learnt, a few years later, that when Idi Amin took over, they were slaughtered for meat as one of the unfortunate consequences of the Ugandan civil war. One of my great sorrows about Uganda is that so many good schemes, which could have made a major difference to the standard of life of the Ugandan people, were aborted because of civil war, political indifference or the unfortunate tendency to discredit any initiative previously instituted by the Protectorate Government. This was done merely on the grounds that the projects were creatures of British colonialism and as such could not be considered of any use to Africans.

In Trinidad the pace of committee work was increased. In 1952 I was elected Dean of the Faculty of Agriculture and as such had to attend numerous meetings in Jamaica, which was the centre of operations for the University of the West Indies. I was, as Dean, an *ex officio* member of many Senate Committees, and if I needed more cash or more staff, I had to make my case in person to the relevant folk in Jamaica. The journey from Trinidad to Jamaica, by air on a DC3 Dakota aircraft, sounds idyllic. Stops were made at Tobago, Barbados, Antigua, San Juan and Montego Bay before reaching our destination at Kingston, Jamaica. Here I was met by a colleague and driven up the Blue Mountains to the beautiful University campus at Mona. But the journey took all day and the return trip another day, so attending meetings in Jamaica took three full working days, and if several meetings needed to be attended in succession, more than three days. This journey came to be known as the 'milk run' and I soon got to know all the British West Indian Airways air crews quite well. The worst of this 'milk run' was the stop in Puerto Rico, which was American territory, for here we had to clear both American customs and immigration on every trip. Computers were not known in those days, and the immigration authorities had a very large 'black book', which contained the names of all the 'undesirables' who needed to be kept out of the USA, even if they were merely passing through in transit. I found that 'Wilson' was a common name in the black book and sometimes I would be held up for 20 minutes whilst the immigration official, not one of the brightest of individuals, pored over every 'Wilson' in his large tome. I discovered that other frequent travellers got through quickly by virtue of the fact that they had a magic 'American priority pass' attached to their passports. I enquired how one could acquire one of these useful documents and was told you had to be well known, and thus vouched for, by the American Ambassador of your country of origin. I had only recently joined the Port of Spain, Trinidad Rotary Club and found that the American

Ambassador was also a member, so it did not take me too long to get to know him sufficiently well to ask him for the magic 'American pass'. My trips to Jamaica were much more hassle-free thereafter, but I noted that I was still regarded with suspicion by the San Juan immigration officers. They probably remembered me from the days when I had no magic pass and they could hold me up at their pleasure.

As Dean of the Agricultural Faculty I had to preside over faculty meetings. My predecessor, Tom Webster, had had the task of introducing new arrangements for entering students for external degrees of London University, of which we were an overseas University College. In my term of office, the University College was transformed into a fully independent University, so it fell to my lot to rewrite both the syllabus and the regulations to make them more relevant to the West Indies. This was a complex undertaking but one which went well, and apart from a lot of 'i' dottings and 't' crossings by the University Senate, we had no difficulty in getting our proposals accepted. At each stage, however, I had to take the 'milk run' to Jamaica to pass the proposals through the various Senate committees.

My most frustrating Faculty meeting took place just before I left Trinidad in 1964. We were ratifying the examination results from the various examiners' meetings. I used the technique of working 'from the bottom up', which meant that we first agreed the names of those who had failed the exam. There was only one name below the line and he had failed abysmally. When I enquired whether the Faculty was willing to fail this student, one member raised an objection. 'What are your grounds?' I enquired. 'The candidate is my cousin!' was the reply. When I asked what that had got to do with it, I was told that if his cousin were to fail, he would lose face in his community and life would be intolerable. We had to debate the issue at some length before I could take a vote and obtain acceptance of the examiners' recommendation. I felt glad I was shortly to leave Trinidad. It seemed to me that, in a few years' time, the examiners would find their recommendations were being reversed and I would not have wished to have been around in those circumstances.

However, my most difficult committee job in Trinidad was not running the Faculty but chairing the Staff Housing Committee. This committee was set up by the Principal to oversee the allocation of houses to staff, and to prioritise repairs and renewals of both the fabric and the furniture. Bunny and I had been accustomed to taking whatever was going and not making too much fuss. We had lived in all sorts of accommodation in Uganda – including periods in African mud huts – and to our mind, all the accommodation in Trinidad was more than adequate for the purpose. Not so, many of the staff – or their wives! When a desirable house became vacant, there was a long list of folk wanting to move in. Whenever there was some money in the kitty for 'refurbishment', there was another long list of supplicants. The job entailed a constant battle with disaffected staff and even more disappointed staff wives. I was delighted in due course to pass over the reins of the Housing Committee to another senior colleague.

I first joined the Rotary movement when I was 33 years old, in 1961. I was proposed by an eminent Trinidadian lawyer of African descent. I had got to know him well as the owner of a plantation in the north-east corner of Trinidad. Because of this he was a regular attender at meetings of the Trinidad and Tobago Agricultural Society, at which I was a frequent speaker. He invited me to visit his plantation on several occasions. At the end of one of these visits he asked me whether I had ever thought of joining the Rotary Movement. I replied that I thought Rotarians tended to be older than I and that younger folk joined the Round Table. I was informed that Trinidad followed the American tradition of Rotary and that Round Tables were only found in the UK. If I wished to join Rotary, he would be pleased to propose me and my age would not prove to be a problem.

So in due course I became a member of the Port of Spain Rotary Club. We met at one of the top hotels in Port of Spain, overlooking the Queen's Park Savannah and our weekly lunches were splendid, leisurely affairs, normally lasting a couple of hours. There was a tradition of having a 'Sergeant-at-Arms', whose function was to fine Rotarians for 'misdemeanours', such as not wearing the Rotary badge, having had one's name in the papers, or having appeared on radio or TV. Unfortunately I ran a daily 'agricultural spot' on Trinidad radio and I was the producer of a weekly TV programme called 'Time to Talk'. Bunny also produced our second son! I was thus picked out for more than my fair share of fines, which had to be taken up to the President's table and placed inside a hollow cricket bat. The proceeds were used to finance the purchase and care of cricket pitches in the poorer villages, since cricket was the national sport and every West Indian child needed appropriate facilities and training.

I left Trinidad after only three years' membership of the Rotary Club, but the Secretary asked where I would be moving to on my return to the UK. When I told him we were hoping to live near Bedford, he wrote to the Secretary of the Bedford Club, so that within a week of our return I was invited to join. Transfer of Rotary membership is not automatic; the theory is that each club fills its ranks with selected individuals representative of the trades and professions of the town or city in question. However, because there is a constant 'turn-over' of members, Clubs are always on the lookout for new recruits and a letter of recommendation from a former Club normally ensures that an invitation to join the new one is not long in coming.

Having successfully avoided committee work in the Port of Spain Rotary Club, in Bedford I was made Chairman of its Community Service Committee, which entailed thinking up a project for the Club to organise. I noted that there was a large local community of Italians in Bedford, who had recently arrived to take up jobs in the many brickworks in the area. British workers, it seemed, no longer wanted to do the hard menial work of filling and emptying the brick kilns. These Italian workers spoke little English (and their wives even less) and often had to communicate with the locals with the help of their children, who were taught in English at school. I decided to recommend that we embarked on a project of helping the Italians to integrate into Bedford social life and we started an Italian Club and also language classes for the Italian women. The projects were successful and attracted the attention of the top brass at the Rotary Headquarters. A year later the Bedford Club was awarded a 'Special Achievement Trophy' for their contribution to international relations! By then I had moved on to Liverpool and joined the Liverpool Club, so I had the pleasure of revisiting the Bedford Club at the time of the presentation of the award by the President of RIBI.

My transfer to the Liverpool Club also went smoothly and after one year as a new member, I was made the Speaker Finder – quite an onerous task, which meant finding someone to talk to the club every week. In theory, I was Chairman of the 'Speakers Committee', but I was told that, although this gave me a seat on Council, I was expected to work single-handedly and that no previous Speaker Finder had ever convened a Committee! How I wish that other bodies I belonged to took a similar attitude to unnecessary committee work! It is truly said that the best committees have only one member!

After three short years in the Liverpool Club, I was moved to Basingstoke. Not only was the courtesy letter written to the local Club Secretary by his Liverpool counterpart, but by chance my brother Donald was a member of the Basingstoke Club and insisted on proposing me. I remained a member of this Rotary Club for over twelve years and served the Club in various ways, as a Council member, Chairman of the International Committee and also of the Community Service Committee. In this latter post I was responsible, with others, for setting up the first Abbeyfield Home for elderly folk in Basingstoke. This was a very worthwhile project and was a great help in making Rotary relevant to the needs of the local community. The Basingstoke Club was much smaller than Liverpool and we got to know the

Rotarians well. One of the leading members of the Club was Ken Harris, a retired senior member of the Electricity Board. Ken was also the Chairman of the Board of Governors of a Secondary School in Basingstoke and he invited me to become a Governor. This was a very interesting experience and I learned a great deal about how the State Schools were run and the differences that existed between them and the Public School system. I referred to some of the lessons I learned during this period in the previous chapter. After I had moved to Edinburgh, the Basingstoke Club celebrated its 50th Anniversary and kindly invited me back to propose the toast to the Club on that important occasion. With a membership of only about 40, it was easy for everyone at Basingstoke to know each other well. Thus it was a great pleasure for Bunny and me to return to see our old friends and to find that we had not been forgotten.

In 1984 we moved back to Edinburgh, but although I was soon invited to join the Rotary Club, I declined. My duties as Head of the School of Agriculture made it impossible for me to absent myself for lunch for two

45. Edinburgh Rotary Club Council members, 1992. Seated, from left: Dr Philip Harris; Mr Ritchie Campbell; Mr Sandy Darling; District Governor John Denholm; Mr Alec Currie. Standing, from left: Mr Keith Frost; Mrs Denholm; the author; Mr John Small; Mr Fred Ainslie.

hours every week and it would have set a bad example to my staff. I therefore let my membership lapse and it was only when I retired in 1990 that I let it be known I would be interested in rejoining, if so invited. One of my two sponsors was John Bethell, then Director of the Scottish Seed Potato Development Council, whom I knew well through my work. He was very much occupied, together with George Holmes, a former Director-General of the Forestry Commission whom I also knew, in promoting the first Rotary Club of the newly independent Ukraine in the capital city, Kiev. This was a pretty major undertaking, especially when John and George persuaded me to become the project manager, which meant a lot of difficult arm's length liaison with two Rotary Clubs in Canada and three more in the USA.

George was the Chairman of the International Committee at that time and he took charge of the Rotary politics by dealing with the officials of the Rotary movement. We organised a major meeting in Edinburgh, to which representatives of the other five supporting clubs were invited. I in turn had to visit the USA once and Canada twice, to pursue the collaborative arrangements and to help with the necessary fund-raising campaign on which the project was dependent. This detailed planning took three years to complete and at the end I organised a delegation of twelve of our members to attend the inaugural celebrations of the Kiev Club. Our Edinburgh Rotary Vice President at that time was Sandy Darling, who not only led our delegation, but also took the trouble to have his speech translated into Ukrainian. All the other Heads of Delegations gave their speeches in English and these had to be interpreted, but Sandy's Ukrainian speech needed no such treatment; it brought the house down and could not have been better delivered or more enthusiastically received.

One of the members of our delegation was Dennis Townhill, a distinguished musician and master of the music at St. Mary's Cathedral, Edinburgh. He offered to give an organ recital during the course of the Kiev Club's inauguration and we were all very disappointed to be told that this could not be fitted into the programme as there was no organ in the hall where the ceremony was to take place. However, we were informed that a separate concert hall, with an organ, had been booked for the purpose, even though the event fell outside the official arrangements for the inauguration. We were more concerned when we learnt that the hall in question was a converted Orthodox Church, some distance from the other events. We need not have worried. On the evening of Dennis Townhill's recital the 'hall' was packed out and loud speakers relayed the recital to the crowds outside. Dennis is a polished performer and his recital was extremely well received.

Dennis ended the recital with a rendering of one of Ukraine's most classic patriotic works, the *Great Gate of Kiev.* The applause after the rendering of this well-chosen piece was something I will never forget. The contributions which both Sandy Darling and Dennis Townhill made did much to enhance the good standing of Edinburgh city and its Rotary Club in the Ukraine and the friendships forged as a result are still active.

There was a moment in our journey to Kiev when I thought that we would never get Dennis to the meeting. When we arrived at Kiev airport, all our passports were called up for examination by the immigration people. Dennis and his wife Mabel were travelling on a single passport. Although this was quite legal, it was in fact a major mistake, as the Ukrainian officials had never before come across two adults travelling on only one passport. They had little or no English and, apart from Sandy Darling's prepared speech, we had no Ukrainian speakers in our party. The airport emptied but our party was still there. 'Two people, one passport. Not possible!' was what the immigration chief was clearly saying. Eventually we persuaded the immigration folk to ring up their head office, which was still in Moscow at that time. Fortunately Moscow had come across the strange British habit of husband and wife travelling on one passport and eventually fax messages arrived instructing the Kiev officials to let us pass through.

The Kiev project did not end with the celebrations marking the inauguration of the new Club. Many Kiev folk came to Edinburgh for 'work experience' and the local Rotarians were marvellous not only at finding them suitable placements, but in offering them home hospitality during their time amongst us. Two years later, we arranged for a party of 'Kiev kids' to visit Edinburgh and spend a week at the Scout Camp at Bonaly. The Edinburgh Rotarians responded magnificently, greeting the children on their arrival, cooking, caring for them and entertaining them during the week, and generally making their stay as enjoyable as possible. There were the inevitable hitches. On a shopping expedition in the centre of Edinburgh, two children disappeared. They were eventually found in the basement of a large store, completely overwhelmed by the range of goods for sale. When the Kiev party were due to come up to Edinburgh by train, there was a rail strike and we had to arrange for the London scouts, who had met the children at the airport and looked after them for the night, to entertain them for an extra 24 hours until the strike was over. Inevitably, on the day when the party was due to go back to London for their return flight home, there was another one-day strike, forcing us to make alternative arrangements to get the party to London by coach. However, the children were not aware of these hitches and the whole visit was much appreciated by all concerned.

46. Visit of 24 Ukrainian children from Kiev to Edinburgh, 1994. The children came from the area near Chernobyl, and the visit was a consequence of Edinburgh's initiative in founding the first Rotary Club in the Ukraine in 1992.

In 1972 my friend from school days, John Sclanders (see also chapter 18), was asked by the Chairman of the Frank Parkinson Agricultural Trust to find a new Trustee with an agricultural background. John thought of me and asked if I would be interested, pointing out that it would first be necessary for me to meet Arthur Parkinson (son of the Founder, who was then Chairman) and his fellow Trustees prior to being formally appointed. The Trust had been started just after the War to award grants of money 'To benefit British Agriculture' and during the past 44 years has provided, in today's monetary values, about £9M in Grant Aid to assist more than two hundred different projects. It operated on the basis of a 'pump-priming principle', expecting recipients to match the Parkinson grant with money from other sources.

When John had made the necessary arrangements for my meeting with Arthur, I duly turned up at a Guildford Hotel and was informally interviewed and invited to lunch at his home just outside the city. The most difficult part of this vetting process was being quizzed by Arthur's wife Marjorie, who was clearly unsympathetic to agricultural matters in general and to intensive livestock production in particular. I can still remember my surprise when she asked me why sheep could not be given waterproof coats

47. The 'Wilson Committee', 1990. From left: Duncan Spring; Tom Kelly; the author (chairman); Frank Armitage (committee secretary); Chris Bourchier; David Leaver.

during the winter to keep out the cold and the wet. When I replied that sheep already had very efficient fleeces for the purpose, she grunted unsympathetically. I have described these several meetings in my book, *A Tale of Two Trusts* (184), so will not repeat the story here. Suffice to say that in a short space of time I was made Chairman of the Parkinson Agricultural Trust and remained in that position until handing over office to Professor David Leaver in 1999, having myself then passed the age of 70. Most of the previous Chairmen of the Trust had died in office and before that had continued to occupy the chair when they were finding it increasingly difficult to carry out their duties. I did not want the same to happen in my case.

The second Parkinson Trust is the Yorkshire Trust, set up in 1950 with the prime purpose of providing Old People's Homes for former workers in the Crompton Parkinson electrical works in Guiseley, Yorkshire. The number of Homes has been greatly increased over the years since 1950, and the present group now comprises 44 units of accommodation, providing old folk with self-contained flats under the overall supervision of a Warden. There are now few remaining former employees of Frank Parkinson's engineering business, so the Homes are open to people from Guiseley in particular and nearby

48. The Trustees of the Frank Parkinson Yorkshire Trust.
Seated, from left: The author, Bill Hudson (Chairman), David Leaver.
Standing, from left: Charles Clough; John Sclanders; Stuart Robb.

parts of Yorkshire in general. The two Trusts used to be run separately, with different boards of Trustees, Arthur Parkinson being one of the few people sitting on both Boards. When I became Chairman of the Agricultural Trust, I thought there would be benefit in linking the two Trusts more closely together. I put this point to the Chairman of the Yorkshire Trust, Bill Hudson, and he readily agreed. The present arrangement is that the Chairman of each Trust sits on the Board of Trustees of the sister Trust, so that there is a firm connection between them. It is very important to find the right people to look after the financial and secretarial aspects of the work of Charitable Trusts. In John Sclanders, Stuart Robb and Janet Smith, both the Parkinson Trusts have been richly blessed and well served by three most able and loyal Secretaries.

Annual meetings are held in London each year, and in addition the Trustees of the Yorkshire Trust meet in Guiseley in September. Visits are made to the Homes, both to meet some of the residents and to have a conjoint meeting with the local Board of Management. I have much enjoyed my connection with these two Trusts and had much pleasure in doing the necessary research to enable me to write up their history. Although both Trusts are modest, each has made a real impact, with many agricultural buildings and extensions and the 44 Homes at Guiseley being a lasting memorial to the memory of their joint founder, Frank Parkinson.

I have also been connected with several other charitable Trusts, all with the common purpose of making annual awards to honour people who have rendered outstanding service to the cause of British Agriculture. These four bodies are the Royal Agricultural Society of England (RASE) Awards Committee, the Massey Ferguson Awards Committee, the Farmers' Club Charitable Trust and the Nuffield Agricultural Trust. These four Trusts all work in roughly similar fashion, although the type of award offered differs in each case. The Farmers' Club award takes the form of a number of bursaries awarded to agricultural teachers. Its purpose is to enable them to undertake a study trip abroad so as to investigate a specific subject and to write up a full report of their findings, which is then published in the Club's *Journal*. I have been Chairman of the Interviewing Panel of the Farmers' Club Trust for ten years and members of the other committees for about the same period and have found the work very stimulating and rewarding. The capital sum, the interest of which pays for the Farmers' Club bursaries, was set up by Mr Muddiman, a senior member of the Club with great ability to obtain money for good causes from his many friends. It was natural that 'Muddy' (as he was known) should be a member of the Interviewing Panel and he always took his duties very seriously. In the mid-1990's Muddy suffered a stroke from which he never fully recovered, but he still came to the Panel meetings together with his wife, who translated his comments on each applicant to the panel. When, sadly, Muddy died a few years later, Mrs Muddiman agreed to continue coming to the meetings. Her knowledge of, and interest in, this Bursary scheme is a great inspiration and we greatly welcome her presence at our meetings. The other panel members are the serving Chairman and Past Chairman of the Farmers' Club, together with one or two representatives of the Agricultural Colleges at which the applicants are employed. They form a very good team. Previous winners of these awards have achieved eminence in their respective fields and the Farmers' Club and Nuffield scholars, in particular, have proved to be very fine ambassadors for British Agriculture and in great demand for lecturing to agricultural clubs and societies.

The Farmers' Club also awards a Gold Cup each year for outstanding services to British Agriculture. I have had nothing to do with this award, but was very delighted when I was presented with this Cup at the Annual Dinner of the Farmers' Club in 1998. One isn't allowed to hang onto the cup for too long as it is very valuable. No sooner, therefore had I received the

49. The author receiving the Farmers Club Gold Cup from Lord Plumb,
Grosvenor House, 1998.

award from the hands of Lord Plumb and returned to my dinner table, than
up came two strong chaps to relieve me of the lump of precious metal and
return it to the vaults of a nearby bank.

I described in chapter 11 my work with the feed trade association, and
mentioned that I chaired its Scientific Committee for over a decade. During
this time I got to know a large number of MAFF officers, both technical and
administrative, who were responsible for framing UK feeding-stuffs
legislation and policing the regulations when they became enshrined in UK
law. I presume this was how my name came up for consideration by the two
Ministers concerned, Health and Agriculture, for a seat on the Medicines
Commission. I duly received a letter inviting me to become a Commissioner
and I served a term of office on this body from 1976-1979.

The matters under consideration by the Commission were confidential
and covered by the Official Secrets Act, which prevents me from describing
any of the details of the work. However, this post gave me some inside
experience of how such bodies operate and the experience I gained over
those three years may be of interest. The Medicines Commission was, and
still is, an appellate body which hears appeals by parties – normally they are
companies but they could be private individuals – who have had their
request for a 'Product Licence' refused. There were two bodies responsible for
awarding, or refusing, these Product Licences: the Veterinary Products

50. Opening of the Information Technology Centre at the Royal Agricultural College, Cirencester by HRH The Duke of York, 1999, partly financed by the Frank Parkinson Trust. From left: HRH The Duke of York; Bunny Wilson; the author; Barry Dent, Principal, Royal Agricultural College.

Committee, which dealt with animal preparations, and the Committee for the Safety of Medicines, which dealt with human medicines. It was no easy task obtaining a Product Licence. A large dossier of information had to be submitted in support of the application and each product had to satisfy the three tests of safety, efficacy and quality.

The civil servants on these two subordinate Committees were very hardworking and conscientious and where the Product Licence was refused, there were normally good reasons for rejection. In most cases, therefore, when appeals were held, the decisions of the licensing authority were upheld. Justice, however, had to be seen to be done and not simply done, so these appeals were taken very seriously. On a few occasions the earlier decision was amended, so that the Licence could be awarded if such-and-such was done to make the product safe. On one occasion an appeal was made by a private individual who had his request for a licence to market his 'magic potion' refused, and for good reason. No dossier was forthcoming showing that it passed the required tests and extravagant medicinal claims were made of the magic potion, along the lines that it could 'cure all ills'. The appellant duly appeared before us and the Chairman very gently asked him to give the basis of his appeal. The petitioner, obviously out of his depth, said: 'I am simply an honest businessman wanting to make an honest living. I wish to go on selling my medicines. But if you learned gentlemen wish to give me advice as to what to add to, or remove from my potions, I shall be only too happy to comply!' Such candour was quite rare and it was with some regret that the Commission upheld the view of the Licensing Authority.

There were about twenty or so Commissioners, but the number of people attending each meeting was at least double this. The 'extras' were the Chairmen of the two subordinate licensing Committees and a whole raft of civil servants, both technical and administrative. Although the Chairman, Sir John Butterfield, was very firmly in charge, he naturally had to take advice from the appropriate civil servants on matters of detail. But John Butterfield had a trick up his sleeve. He spent much time getting to grips with the minutiae of just one item on the Agenda, but no one knew in advance which this was. When the item came up, Sir John turned to the appropriate official and asked for a summary of the case. The official, cherishing his ten minutes of glory, would hold forth on all aspects of the case and eventually arrive at his recommendation. The advice would appear to be impeccable, as it was full of detailed references to the EEC Directives and Regulations and the UK Laws and Statutes which laid down the procedures to be followed. Sir John would listen until the civil servant had finished and would then say something like, 'I note you have quoted Statute 6 chapter 9 paragraph 10(b) and if that was all there was on this point, your advice would be correct. But there would appear to be a contradiction with Statute 5, chapter 6 and paragraph 12(c). How do we get over this little difficulty?' Exit baffled civil servant weighed down by a mass of files, to go back and do his homework more thoroughly!

In 1976 we had only been members of the EC for a few years. As a result, few people knew the extent to which EC involvement would affect our work, or which UK Laws would have to be amended to bring them into line with European legislation. The Medicines Commission had agendas which were mostly concerned with hearing appeals. There was little time left for lateral thinking. Now John Butterfield was keen that we should take a step back and debate the ways in which European membership would in future govern our decision-making processes. The civil servants were not keen on this and came up with the argument that it was not a proper part of our statutory business. But John was not to be put down. 'I know there is no time at our meetings,' he said, 'so I propose we have a weekend retreat. You would all be very welcome to debate these issues in Cambridge.' (John was then a Master of a Cambridge College). The civil servants were appalled. Not only would we be discussing delicate matters of a political nature, but they would have to give up a weekend for this purpose. 'Don't worry,' said John, 'I only want the Commissioners to come. Civil servants are excused!' Consternation all round. Commissioners meeting in secret without civil servants to keep them in check! One of the civil servants played what he thought was his trump card: 'If you meet informally in the way you suggest, we cannot regard this as a properly constituted meeting of the Commission, *and we won't pay expenses.*' 'That's not a problem,' said Sir John, 'They will all be my personal guests at College and no expenses will be incurred except the costs of travel to Cambridge. I am sure my Commissioners will be able to get there under their own steam!'. Game, set and match! We met at Cambridge, where we had a useful, as well as an enjoyable, weekend. Whether or not the exercise was ever repeated I do not know. It was the only retreat during my time on the Medicines Commission.

I have already described my introduction to committee work with the Youth Fellowship at Sanderstead and the Youth Club at Wye, where the chairmen were the Vicars of the parish churches. In the years following I rarely spent long enough anywhere to become involved with parish administration, but I continued to attend church and did what I could. In Uganda the Bishop of our up-country diocese, 'The Diocese of the Upper Nile', regularly stayed with us at Serere when visiting the area to officiate at services. Once he invited us to visit him at his home and showed us round his cathedral. He described how it had once been taken over by a quasi-Christian religious sect. When asked what they were doing there, their leader explained that he had had a vision in which God had told him to come and live in the cathedral with all the members of his sect. 'When did God tell you this?' the Bishop enquired. 'Two weeks ago,' he replied. The Bishop returned to the cathedral early the next morning to tell them that he also had just had a vision in which God had instructed him to tell the sect that it was time for them to leave! They did so without further argument.

Our relationship with the Bishop of Trinidad was less intimate, but when a small mission church near the University lost its incumbent, I was asked if I would conduct the weekly Matins service for the mainly Asian congregation. I protested that I was not qualified to do this, but the Bishop promptly licensed me as a Lay Reader and introduced me to my congregation. For several months I did my best to take the weekly services and was warmly welcomed by the local people. I think this temporary arrangement might have continued but we were leaving Trinidad. At my last service I was most surprised and touched to find that the congregation had arranged a presentation, thanking me for what I had done for them. It was a very humbling experience to find the little I had given had apparently meant so much.

On our return to the UK we attended the local parish church in Bedford. I took over the running of the Sunday school for a couple of years and started a Youth Club, but being then 35 years of age, I found that I did not have the affinity with youth as I had had when aged 21. Their interests had become more sophisticated over the years. The most popular activity was a loud disco, the success of which seemed to be directly related to the number of decibels produced by the loudspeakers. I shared the running of the Youth Club with a colleague from Unilever, Derek Shrimpton, and we found that we had to take it in turns to escape outside the hall every half hour to rest our ear drums. I was on the Parochial Church Council (PCC) and we got to know the Vicar, Peter Mumford, and his family very well. Peter subsequently became Bishop of Truro and was tipped for further preferment but he tragically died whilst still relatively young.

It was not until we settled in Farnham that I became involved in the running of the parish. Both Bunny and I took turns to be on the PCC and I formed a group, known as 'Parish Roundabout', which met for talks, debates and social events at the three different parish centres in rotation. We produced a 'parish film', a collection of colour slides with a taped commentary, which we used when welcoming newcomers to the parish. The meetings, held monthly, were popular and I managed to run the Parish Roundabout with a committee that met twice a year to plan and oversee the programme. We had a succession of excellent Vicars; one went on to be Bishop of Peterborough and another Archdeacon of Dorking. It was a lively parish and we made many good friends.

On our return to Edinburgh, we rejoined Christ Church, the Episcopal church at Morningside, which we had first attended in 1950. It is sited at 'Holy Corner', so called as there are (or were) four different churches on the four sides of a cross-roads. There is quite a lot of interaction and co-operation between the four, especially in the running of an inter-denominational organisation known as the 'Eric Liddell Centre' catering for the needs of various groups, both secular and ecumenical. As at Farnham, Bunny and I took it turns to serve on the Vestry (the Scottish equivalent of the PCC) and twice we were involved in the selection of a new Rector. I also served for a period as the 'Lay Representative', which entailed representing Christ Church on the Area Council and the Diocesan Synod and also serving on the Synod's Finance Committee. I was on the Diocesan Synod at the time the definitive vote was taken on the 'Ordination of

Women'. I spoke in favour of this move, which I considered overdue. I was struck by the fact that most of the opposition to the proposal came from two unexpected quarters, the House of Clergy and some very outspoken women Lay Representatives. The motion was duly carried, but the reaction of these two groups struck me as being very odd.

I found that I was attending Church committees or meetings at least once a week, with the usual frustrations I have described earlier. Rarely did I find the chairmen had any training or aptitude for the job. Sub-committees had either not met or they had not had time to produce a report and it seemed that a meeting which should have finished its business in an hour continued for about four hours. Part of the problem, I discovered, was that some folk regarded these committees as 'social events' and thought of them as 'a night out with friends'. The result was that socialising was given priority over the business. Doubtless I am being somewhat hypercritical, but this sort of attitude is sometimes just a little difficult to accept. The Church seems to take pleasure in doing things in an amateurish fashion and although this helps to bring more folk into the running of Church affairs, unfortunately it does not lead to a more efficient conduct of important business.

In 1990 the Scottish Episcopal Church started involving itself in environmental issues. This interest was entirely justified, for 'husbanding the world in which we live' is an important part of religious ethics. Unfortunately, however. It meant that that the church tended to lean on, and accept too readily, the more extreme utterances of the 'Green Lobby', which were often synonymous with the 'Anti-farming lobby', and I objected. It seemed to me another example of a flirtation with important subjects which the Church had neither the time nor the talents to investigate, with the result that superficial answers to complex questions were being given. To cut a long story short, this eventually resulted in my being asked to sit on a Committee of the 'Science, Religion and Technology' group of the Church of Scotland, which was embarking on a study of the Ethics of Genetic Engineering. This was a very interesting assignment, which lasted six years. The Committee was ably serviced by Donald and Ann Bruce and my fellow members were influential and knowledgeable folk who commanded my respect. From Edinburgh University came Mike Appleby, Reader in Animal Welfare, who had been a valued member of my staff when I held the Chair of Agriculture. David Atkinson, the Deputy Principal of SAC, brought with him a deep knowledge of agricultural research and as a member of the General Synod of the Scottish Episcopal Church, he was able to bridge the gap between science and technology in that organisation. John Eldridge introduced the sociological perspective as Professor of Sociology at Glasgow. Michael

Northcote was a Senior Lecturer in Christian Ethics and Practical Theology at New College, Edinburgh. Joyce Tait was an ecologist and Deputy Director of Research at Scottish Natural Heritage. Ian Wilmut was a senior scientist at the Roslin Institute in Edinburgh and takes the credit for being a senior member of the team which produced 'Dolly the sheep'. Finally Michael Wilson, a plant breeder, was the Deputy Director of the Scottish Crop Research Institute at Dundee.

We were an interesting team, who had to spend quite a lot of time in the early years of the study finding common ground; there were times when I never thought it would be possible to reach agreement on any aspect of our study. Donald Bruce wisely took the view that we should produce a book which mapped out the area, explained our different viewpoints, illustrated the study with case-histories and left the readers to come to their own conclusions on the questions posed. This was a 'winning formula' and a very readable book resulted from our labours called 'Engineering Genesis' (183). My own specific contributions were in helping to draft the chapter on 'Genetic Engineering in Developing Countries' and in producing the case-history on Bovine Somatotropin (BST) (a growth hormone which increases milk yields in cows), on which subject I had previously written a paper (165). The whole study was an excellent example of how the Churches should tackle highly complex subjects and it is to the credit of the Church of Scotland that they conceived and funded this major initiative.

I have always supported the 'ex-students' associations of the institutions that have been instrumental in teaching me throughout my career. Thus I have been a keen member of the Old Whitgiftian Association (OWA), the Wye College Old Students' Association (The Agricola Club) and the Edinburgh Graduate Students' body, known as the General Council. On my retirement, I allowed my name to be put forward by Ian Cunningham (see chapter 18) for election to the Business Committee of the General Council. I served on this Committee for nine years, firstly as a Business Committee member, then as Convener of its Constitutional Standing Committee and latterly as Vice Convener of the Business Committee itself. I have enjoyed the work, although at times I have wondered whether all the effort we put into the job really paid off, in terms of positive help given to the University. However, the General Council is a Statutory Body, enshrined in the Universities of Scotland Act, and it has certain important functions to perform, such as the election of the University Chancellor.

I have only space to recount a small number of the 'Committees and Councils' which I have served on over the years. The experience of playing a part, even if only a minor one, in 'democracy in action' has been interesting

51. Farewell Dinner in honour of Ian MacLaren, Secretary of the General Council of Edinburgh University, 1997. From left: The author; Mrs Margaret Tait; Mr Ian MacLaren; Dr Martin Lowe.

and, at times, stimulating. Much depends on the skill and personality of the chairman. With good chairmen, the task is enjoyable since the members of the committee work together for a common cause as a unified team. With a bad chairman, in-fighting and sectional interests tend to prevail and the real *raison d'etre* of the council or committee tends to get lost sight of. I have served under some really splendid chairmen over the years, but also under a few 'rogues and vagabonds' who have hi-jacked the organisation to further their own interests. I believe that all good chairmen are basically reluctant to take over the reins. They would happily stand down in favour of a better person, should one become available. One should always beware of chairmen who are ambitious to reach high office. Indeed, when one notices such people manipulating their way up the organisational tree, the alarm bells should start to ring. I am pleased that, in my experience, such people are in a minority, but it is nevertheless hard going when one is landed with a bully of a chairman, since some folk are clever at hiding their ambitions from their colleagues until it is too late to stop them reaching positions of power. Perhaps the most efficient committees are committees of one, but then the

*52. Receiving an Honorary Degree of Doctor of the University from Stirling University,
2000. From left: Professor Randolph Richards, Director, Institute of Aquaculture;
Dame Diana Rigg, Chancellor of Stirling University; the author; Principal Andrew
Miller, Vice Chancellor, Stirling University.*

checks and balances which are necessary to curb excesses are not in place
and things can go badly wrong. I often wonder whether the many thousands
of hours I have spent 'in committee' was time well spent, or whether I could
have been more gainfully occupied. '*Oh, give me your pity, I'm on a committee*'
certainly does have a ring of truth about it.

CHAPTER 17

Happy Families

'All happy families resemble each other,
Each unhappy family is unhappy in its own way'
Leo Tolstoy

SOME FOLK POINT OUT that you are able to choose your friends but not your family, and use this as an excuse to loosen family ties and look for relationships elsewhere. Bunny and I have been lucky enough to have been born into caring families and we like to believe that our children have been similarly fortunate. The warmth and security of a close-knit family is of immense value and one can only feel real sorrow for those friends and colleagues whom one encounters, whose family life has been unhappy or even painful. This chapter attempts to describe the known members of our extended family. We have not made a study of genealogy and the history of both our families is mainly reliant on human memory and entries in the large family Bible, now unfortunately lost. But one interesting fact does emerge clearly from these elementary studies. As one goes backwards through the generations, on average one extra person can be added at each step. Thus the demographic trend towards smaller families, at or near replacement rate, is clearly evidenced in our own family history.

The Wilsons

My parents were married in 1913 and produced twin boys a year later. My father served in the Royal Navy as a writer during the First World War and was stationed at Portsmouth and my mother had a very hard time raising two children on a serviceman's wife's allowance through the difficult War years. They lived in rented accommodation but never spoke much about it, although I have the impression that it was pretty basic. After the War they moved into a council house in Goddard Road in Beckenham, Kent and I was born in a local nursing home in 1928. In the years of the depression things were difficult and no doubt my mother had to watch the pennies very carefully. My father, although having a good position after the War with a London firm producing high quality confectionery, was never well paid. Even when he became a Company Secretary and Director during my

53. Our Wedding, September 9th, 1950. Standing, from left: Brother Donald (best man); Edna Tolly; my father; the author; Bunny, Bunny's father, Elizabeth Ford; Joyce Bunn. Seated, from left: my mother; Bunny's mother.

teenage years, his salary never rose above £1,000 a year and for most of his career was far less. The result was that my brothers and I were brought up in a stable environment of a lower-middle class home without many luxuries. A 'big treat' was to have tea in a cafe after going to the 'pictures' in the afternoon, or going on annual holiday to Frinton-on-Sea, as described in the first chapter of this book.

When my father was over 90, he decided to write down that part of his family history which he could remember. The following extract is taken directly from his typed manuscript:

'The line starts with my paternal grandfather, born towards the end of the reign of George III of Scottish parents. He was an officer in a regiment then called the British German Legion, based in Heligoland, long since disbanded. We had, at home, some regimental dinner plates with the regimental crest, a Bugle surmounting the letters B.G.L. After leaving the army, he obtained a post as station-master at Manchester and subsequently became manager of the Army Clothing Depot in Pimlico, which made as many of the military articles required as were within its competence and bought in other lines, such as helmets and boots, which they did not make. Whilst there, he employed his daughter, my Aunt Nell, and his son, my Uncle Arthur. He married a Welsh-speaking lady. His eldest son, James, was my favourite uncle. He was a priest of the Church of England and spent some years in New Zealand, where he met his wife, Mary. He subsequently returned to England, where his last charge was as Vicar of Foxton in 'The Shires'. He had a stroke and ended his days in the Homes of St. Barnabas at Dormans, Surrey, where my wife and I frequently visited him. It is a home for aged and invalid clergy of the Church of England. My Uncle David was a gilder and picture-frame maker and had an Artists' Colourman's business in Eastbourne. His shop was in Terminus Road opposite the station and when he left Eastbourne, he sold the lease to Hudson Brothers. I met my wife in his house. When his family grew up and left home, the marriage broke down. His wife and children migrated to the Americas separately and I have since lost touch with them.'

My father had one full sister, Dorothy, who was born in 1890. She married Oliver Jenkinson and emigrated with their four children to America, but no effort was made to keep in contact with her. Both my father and his sister Dorothy had a most unhappy time (which I have described in chapter 19) as a result of losing their mother when they were young children. My father's favourite uncle, James, had five children: James, Osmund, Irene, Muriel and Philippa, and it was with this branch of my father's family that the ties were strongest. Only the eldest son, James, married and had four children. His sister, Irene, who was born in 1891, had an unusual career for a woman of her time and position. She travelled to Italy before the War, learnt to speak Italian fluently and then got work firstly as an English speaker on an Italian radio programme and then as a senior official in the British Embassy in Rome. When War broke out, she returned to the UK, and because of her diplomatic experience, was appointed to Winston Churchill's secretarial staff. She was also employed by the BBC Overseas Service for their foreign language broadcasts in Italian for some of the War years.

Irene lived to the age of 99 and we saw a lot of her during the last twenty years of her eventful life. She lived in Bungay in Suffolk for many years, but had always wanted to become a resident of the Old Peoples' Homes in Ditchingham managed by the St. Anne's Order of nuns. This Order was also responsible for running the boarding school which Irene had attended as a young girl. When a room became available in the Ditchingham Homes, she moved in and spent the last decade of her life very happily in this religious community. She looked forward to our visits, during which we took her for car rides round the East Anglian countryside which she knew so well. At her request, we also drove her to morning service in her old school chapel, where she was regarded with a great deal of respect as she was for many years the school's most senior Old Girl. Three years before she died, we got a phone call from Irene's doctor. He told us that she was unwell and had 'turned her back to the wall', which meant that she had given up the will to live. We hurriedly bought an air ticket to Norfolk airport, hired a car and rushed down to Ditchingham to pay her our last respects. On rounding the drive to her Home, however, we were amazed to find that she was standing outside the door, wearing her hat and coat, waiting for us. 'Have you come to take me out to lunch?' she asked, 'How nice of you!' The matron apologised profusely, saying, 'When she heard you were coming, the "will to live" returned!' If anyone learned the art of growing old gracefully, it was Irene. Towards the end of her life she was nearly blind – she suffered badly from tunnel vision – and was growing increasingly deaf. Being a sociable soul, she found that these disabilities cut her off from the other old folk in the Home and she grew increasingly lonely. We were sorry not to be able to visit her more often in her last few years, but Edinburgh is a long way from Ditchingham and the journey is not an easy one. Irene was really our last close connection with the older members of my father's side of the family and she was a worthy and much loved member of the Wilson clan.

My mother was born Fanny Louise White in 1889, but I have never heard her use either of these Christian names. Her contemporaries called her by her nick-name, Dolly, but I was never told how that came about. Incidentally she always referred to my father as 'Monty', derived from the last of his four Christian names (Llewellyn William Charles Montgomery). I have already described how my father first met my mother in Eastbourne. Photos show that she was a strikingly pretty young girl, with lovely long dark hair. My mother's parents were Arthur White and Louise Francis and I remember them well as we used to visit them at their home in Mill Road, Eastbourne when I was small. I can remember arriving at Eastbourne on one occasion (I would have been about three at the time) and asking where 'Grandad' was, to be told that he was digging new potatoes for our lunch and I would find him at the far end of the garden. I remember helping him carry the potatoes back to the house, keeping some for cooking and putting the rest in a sack in a large cellar, approached by steep steps from the back garden. My mother's grandmother on the female side was born in George IV's time and lived to her 99th year. The result was that my two brothers remembered her well as a grand and lovely old lady, but she died before I was born and so I have no knowledge of her except what I have been told by my parents. She was, apparently, very close about her family and could not be induced to speak of her childhood or her marriage.

My mother's paternal grandparents had four children, three sons and one daughter. The daughter, my mother's Aunt Lucy, married James Hughes and they had one daughter. This cousin of my mother was baptised Hester, but was always known within the family as 'Lassie' and I remember 'Aunt Lucy and Cousin Lassie' from the days when they lived in Eastbourne in a large house, which I considered very grand. After Uncle James died, Aunt Lucy and Cousin Lassie moved to Acton Town in North London, where they lived in a Victorian semi-detached house. We visited it from time to time and were very hospitably entertained to excellent luncheons, which to me seemed little short of feasts. Aunt Lucy died after the War and Lassie, who never married, continued living in Acton Town on her own. Lassie had had a varied life, starting out as an assistant in a high-class jewellers, then becoming a senior member of the Almoner's Department in Acton Hospital during the War and finally becoming a member of staff of the Victoria and Albert Museum in Central London. A very small and petite woman, who was stunningly pretty in her younger days, Lassie finds life very dull and frustrating now she is confined to her house by poor health. We have done our best to keep up with Lassie over the years, and Bunny especially has been a regular visitor to give what support she can as Lassie has advanced in years, but, as with Irene, the distance has prevented us from keeping as closely in contact with her as we would have wished.

In 1932 my mother's father died, and my grandmother sold her house in Eastbourne and came to live with us in Sanderstead. My grandmother was a lovely lady. She was short but amply endowed, cheerful, loving and cuddly. She was able to show me the warmth that unfortunately my mother could not and lent a very welcome shoulder for me to cry on when I was in trouble. She helped around the house as best she could, but the arrangement of sharing our relatively small home in Sanderstead never worked out. Her only 'private space' was her bedroom and she stayed in it for much of the day so as to get 'out of my mother's way'. She was 'expected' to fall in line with whatever we were doing and any attempts to 'do her own thing' invariably got her into trouble. Although my parents were not teetotallers, alcoholic drinks were only produced when there was something to celebrate. Grandma liked her occasional drink and used to steal away once a week by bus to have a glass of port in one of the Croydon pubs. Somehow my mother got to hear about this and there was a dreadful row. Public houses, according to my mother, were dens of vice and no one should be seen frequenting them – especially elderly unaccompanied ladies! So one of life's few remaining pleasures was taken away from Grandma, and she retired increasingly often to her bedroom to 'read a book'.

My mother's father was buried in Eastbourne and once a year Grandma made a pilgrimage to see his grave. She had paid the authorities a sum of money in order to ensure that the grave was properly tended and kept free from weeds. Grandma invariably came back from these visits very upset. She always found the grave overgrown and neglected and not only felt cheated, but experienced a sense of guilt that she had in some way let her loved-one down. It took her many days to get over these frustrating visits to Eastbourne and I felt very sorry for her and did my best to console her. My mother was less sympathetic, arguing that there was a war on and doubtless all the workmen previously employed to look after the cemetery had been called up. When Grandma died in 1951 and her will was read, my parents were horrified to learn that she had expressed a wish to be cremated. My parents were at that time very traditional and did not think that cremations were fitting ways in which to depart this life, but her wishes were respected. How things change, for when my parents died, both had decreed that they should be cremated – a decision which I could not imagine them taking twenty years previously.

I used to confide more in my grandmother than my mother, whom I always found to be on a different wave-length to me. After I left University, I was closer to my father than my mother, as Dad followed my career with knowledgeable interest. My mother, on the other hand, never quite understood what I was up to, since the sort of work I was doing and the places I was living in were so alien to her limited experience. Perhaps one needs to go back to my mother's childhood to gain some insight into the way her mind worked. Although brought up by caring parents in a pleasant home in Eastbourne, money had never been plentiful and I suspect my mother's own childhood had not been without difficulty. It was expected that children in her station of life should not go to the local state school, but private school fees were beyond the household budget. In order to solve the problem, my mother played the piano at a private school, in return for which the fees were reduced. This must have had the effect of setting her apart from her school contemporaries. She was the 'piano player' and therefore not a 'normal' pupil. Although this secured her a good education, it clearly made the establishment of normal childhood friendships a lot more difficult. Although Mother was a very intelligent and well-read woman, she was never at her best when visiting or entertaining either friends or relations and I found it far easier to arrange my own teenage parties in other friends' homes than to ask my mother to hold them in our house.

An isolated incident illustrates this rather vividly. When my brother Donald was courting, he asked my parents if he could invite his fiancée, Doris, and her parents to share Christmas lunch with us. My mother was a very good cook although she had a limited repertoire. She always excelled herself at Christmas and a perfectly roasted turkey, with all the trimmings, could be confidently assured. Donald thought that the meal would impress Doris and, even more to the point, her parents. So the invitation was duly given for lunch at 12.30 on Christmas Day. The weather was exceptionally bad that winter and all Christmas morning the snow fell heavily. Doris and her parents had to come by bus, which necessitated changes; one bus from Bromley to Croydon and another from Croydon. Inevitably, the buses were delayed and the final leg of the journey had to be completed on foot. All the passengers had to disembark about a mile short of their destination and trudge the rest of the way on foot through the blinding snow. Eventually three very bedraggled and white-coated folk arrived – but at about 2 p.m.! They were greeted by my mother, who exclaimed, 'What a pity you didn't get here on time. We've had our Christmas lunch but please come in and have some mince pies and coffee!' Donald never forgot this incident and it is a wonder that it did not bring a promising romance to an abrupt conclusion. Many years later, when I in turn invited Bunny to Christmas lunch at my home, I said very clearly, 'Whatever you do, for goodness sake don't be late!'

My twin brothers both went to Beckenham Grammar School and duly passed the usual examinations. Many folk could not believe that my two brothers were twins. Although exactly the same height as my father (six feet), and with the characteristic 'Wilson balding pattern', they differed in almost every other way. Arthur was the more heavily built and stronger twin, good at sports, down to earth and pragmatic in his attitude. He remained a staunch churchman all his life and was a leading member of his small Cornwall parish church when he died. Donald, on the other hand, was slim and angular. He was much more sensitive and artistic in outlook and more questioning in his attitudes to such matters as religion and politics. When visiting Donald, one had to be prepared for a string of questions on current affairs and matters of topical debate and he did not take kindly to sloppy, ill thought-through answers. To my knowledge, he never played any outdoor games and his main physical recreation was walking, which he did quickly (in common with most male members of the Wilson family) at something approaching Guards' pace. Donald had been very ill as a teenager and missed school for almost a year because of a bad attack of pneumonia. I believe my mother nursed him at home during most of this period, but he spent some considerable time in hospital in the early stages of his illness.

54. My brother Arthur, Palestine, 1941.

Although our home was a caring one and certainly provided all the basic necessities of life, it appears that the environment was somewhat inhibiting to two teenagers wishing to find their own feet. They were, however, both involved with the scouts, becoming King's Scouts and eventually joint Scoutmasters of their troop. This meant that they were away several evenings each week and at most weekends. I did not encounter a similar problem, as I left home to go to University at the age of 17. In any case, I seem to have been given more freedom than my two brothers. Donald would have liked to have trained to be an architect. However, he received no encouragement and left school after taking his School Certificate exams. He earned his keep by working as a demonstrator/salesman in a Gas Board Showroom in Croydon, work he never enjoyed. Arthur, on leaving school at the same time, worked for an insurance office in London, but the company was not a well established one and when he brought the annual accounts of his firm home for my father to examine, he was quietly told that the company was about to go bankrupt and that he should look for some safer position. This advice was taken, but not in the way my mother and father had imagined. Instead of finding another job with a 'safer' company, Arthur informed his parents that he had signed up for the Palestine Police and would be sailing in two weeks' time. My father took the news in his usual pragmatic fashion, but my mother was hysterical and did all she could to reverse Arthur's decision. However, it was all to no avail. Two weeks later Arthur duly caught a boat from Southampton and we did not see him again until several years later. Donald joined the Special Police as the War approached, and spent many evenings away from home. When war broke out he was in a 'reserved occupation' for a time because of his police duties, but he was eventually called up and joined the RAF. One way and another, I saw very little of either of my brothers during my formative teen-age years.

After a long stint in Palestine, during which period Arthur was lucky to escape being killed when the King David Hotel in Jerusalem was blown up by terrorists, he transferred on promotion to the Uganda Police. He went out as an Inspector and retired as Deputy Commissioner of Police, having been awarded the Colonial Police Medal for his services. Arthur was a bachelor throughout his period in Palestine and also for his first tour of duty in Uganda, but on one of his home leaves his attention was drawn to the pretty girl who lived (literally) next door, Joan Hunter. Soon afterwards they were duly married in our parish church at Sanderstead. I was an usher at their wedding, but I missed the marriage of my other brother, Donald, to Doris Goringe a year or so previously, as I was then in hospital recovering from scarlet fever. I was very sorry to miss this wedding as I was very fond of Doris, who had been a very good friend to me. When Donald was in the RAF, I used to cycle out to her home in Bromley, Kent about once a month and she would provide me with a splendid tea with lots of home-made cakes. We also had long 'adult' conversations on subjects that never got discussed at home. Not to be outdone, on one of his short leaves from the services, Donald took me to a restaurant in Croydon and gave me tea. Knowing my liking for meringues, he said that I could have as many of these goodies as I could eat. I surprised him when I managed to consume no less than fourteen. After the War, Don had several jobs but later joined the German-owned company, Wella, which specialised in the manufacture of ladies' hairdressing equipment. He was eventually promoted to Sales Director of the company and moved house to Basingstoke in Hampshire, where Wella had its UK headquarters.

Bunny and I caught up with Arthur and Joan when we went to Uganda in 1950, and we joined up again with Don and Doris in 1971, when I was moved by Unilever to its Basingstoke office. On leaving Uganda, Arthur moved to Crowborough in Sussex, where he got a job, firstly as an insurance salesman for Canada Sun Life and latterly a way-leave officer for the Post Office. On final retirement, Joan and Arthur went to live in a pretty cottage at Ruan Highlanes in Cornwall, whilst Don and Doris left Basingstoke to retire to a house in the village of Childe Oakford in Dorset. It follows that my two brothers saw very little of each other after they left home, although contact was renewed when Arthur retired from Uganda. Although 14 years their junior, I saw more of both of them when we lived in Farnham, until we returned to Edinburgh in 1984.

Don and Doris had no children, but Joan and Arthur had a family of two, both born in Uganda, which subsequently led to passport problems when in turn their own children were born in Germany. Clive was born in 1946 and Valerie in 1954. Clive went to Tonbridge School and thence to Sandhurst, where he was duly commissioned at a passing-out parade presided over by Field Marshal Montgomery. After many postings, including at least two to Northern Ireland, he left the army on one of the 'early release' schemes and took a degree in Art at London University, then becoming a management trainee at John Lewis' in London. However, his artistic talents and his wish to carve a more independent career path led him to leave paid employment and start on his own account as a consultant building contractor. Clive has done exceedingly well in this specialised niche market. Clive married Chantal Lavotaire and they have two girls, Eleanor and Louise. Both their daughters are currently completing their university education. Valerie married a banker, David Rose, and they have two children, Gabrielle, born in 1983, and Chantal, born in 1986. Both their girls are very bright and enjoying secondary school at the time this book is being written. They are also good at chess and play competitive chess at junior county level. My generation of the Wilson family has remained close, even though we are pretty widely scattered and opportunities for large reunions are consequently difficult to arrange. This is a very different state of affairs to that of my father's family, which was unhappy in many ways, some of which I have described here and also in chapter 17.

The Bunns
Bunny was an only child. Her parents married in 1925 and she was born in

55. Bunny and three generations of the Wilson clan, 1980.
Standing, from left: the author; my father; Bunny; my son John.

1926. Her father, William Bunn, was born in 1901 and was the eldest of four
children, The others were Maud, who married George Morgan but who had
no family; Harry, who died unmarried when he was only 19, and Leonard,
who married Alice Newton and had three children. Bunny's father, William,
was a short, stocky man with a most engaging smile and cheerful manner. He
was extremely intelligent and would have greatly benefited by going on to
further education from school, but he had to leave at 14 to supplement the
family income. He started work in a cigarette factory, where he met his
future wife. The story goes that he decided that he would be better off as a
lorry driver. This was in the days before driving tests were necessary, so he

persuaded a friend to take him out a couple of times in order to teach him to drive. On the strength of this minuscule apprenticeship into the skills of lorry driving (and this was in the days when double-declutching was necessary), he responded to an advertisement, 'Lorry Driver Wanted' – and got the job! However, lorry driving did not stretch his skills fully, and when I knew him he was the foreman/manager of a company which made aeroplane seats. With this background and his inherent intellectual ability, he encouraged Bunny throughout her progression from school to University and took great pride in her achievements. He was in this respect like my own father and they got on very well together, although their own family backgrounds were very different. Certainly Bunny's mother and father had a happier childhood than my parents did, even though they both came from a more modest background. Both our fathers chain-smoked, but my father always smoked a pipe and did not inhale, so that his lungs did not suffer as a consequence of his habit. Bunny's father, on the other hand, smoked cigarettes and inevitably this was the cause of his early death at the age of 61.

When I first met Bunny in 1949, her maternal grandparents were both dead but her paternal grandmother, Laura, was still alive. Laura was, in many ways, very similar to my own grandmother. She was small, cheerful and devoted to her family in every way. She lived in a flat in north London when I knew her and survived solely on her old age pension. Her husband had spent several years in South Africa working in the dockyards at Simonstown, but Laura had refused to go with him and had lived alone with her children for the years he was away. The background to this separation is not fully understood. They were a devoted couple but apparently Laura refused to leave her widowed mother. Perhaps too, she feared a long sea voyage. I always thought her very fortunate in that her husband did eventually return home, although apparently not much richer for his African adventure.

Aunt Maud was a great favourite, very fond of her mother, Laura, and very like her in many ways. Having no family of her own, Aunt Maud (who was an accomplished needle-woman) spent many hours making clothes for Bunny when she was a young girl and also took her on holiday. Maud's husband, George, had been a stoker in the Royal Navy during the First World War and joined the London Police after being demobbed. He had damaged his lungs through shovelling coal in the ships' boiler rooms and at some stage had injured his leg badly. As this was in the days before antibiotics, the wound never healed properly and he was troubled with a lame leg for the rest of his life. He always refused promotion in the police as that would have involved a medical examination, which he feared he might fail, thus bringing his police career to an unwelcome end. Originally living in North London like the rest of the Bunn family, in the 1960s the couple moved to Rustington-on-Sea in Sussex. Although only a mile from the coast, Maud and George spent most of their time maintaining their bungalow and garden, which was always in immaculate condition. It seemed that the only occasions when they walked the promenade and enjoyed the sea air were when we took them there by car. Having no children on which to lavish her love and affection, Maud spoiled George in every way. On our various visits to Rustington, we used to take them out for a good meal and they both enjoyed these trips, as they did very little eating-out on their own. It used to amuse me, however, as Maud always took charge of the menu and told her husband exactly what he would like to eat! But George didn't object in the slightest and clearly enjoyed whatever dish was set before him.

Bunny's uncle, Len, worked in the office of an engineering company, Browns, and his three children were Joyce, born in 1941, Len, born in 1943 and Gerald, in 1954. Joyce married but like her Aunt Maud, had no children of her own whilst both the boys married and between them had five children. Joyce married John Chapman in 1964, Len wedded his Norwegian girl friend, Siri, in 1965 and Gerald married Gill in 1977. All these marriages took place after we returned from Trinidad and so we were able to attend them all and also, in turn, the marriage of Len's eldest daughter, Ingrid. When we lived in Farnham and had the advantage of a large house and garden, all Bunny's side of the family used to visit us for big family reunions, known as 'Bunn fights'. These were very happy occasions and somehow or other they were always blessed with fine weather, so that the younger generation could play happily together in the garden whilst their parents sat down with a long cool drink and reminisced. We all look back on those days with great pleasure. They were good days and were over far too soon.

Bunny's mother, Ethel, also left school at 14 and was apprenticed to a tailor, but having been employed sewing on buttons for a year and receiving derisory wages, she gave it up for work in a cigarette factory, where she met Bunny's father. Bunny's parents had a double wedding with Maud and George. Bunny (baptised Maud after her aunt, but always known as 'Bunny') was born a year later and grew up in a close and supportive family. In those days it was unusual for married women to work and Ethel's days were spent keeping her home spick and span. This continued until the War, when for a short time she was drafted into 'War work'. She had never enjoyed good health and was soon exempted from this War-time occupation on the advice of her doctor. She was devoted to her daughter and no one had a more caring mother. When I was courting Bunny and visited her home in Wembley, I found Bunny's parents very hospitable and they welcomed me

56. *Ruby Wedding celebrations, Centre Park, Nottingham, September 1990.*
From left, at back: Rosemary; Andrew; David Lewis; Bunny; Irene; David Wilson; the
author; Carol Wilson. Children standing in front row: Stephen; Michael; Julia; James.

into their house unreservedly. They had a large brown dog called Ted, of
indeterminate breed, to which they were very attached. I believe it was
necessary for me to pass the 'Ted Test' to be fully accepted into the Bunn
household, but fortunately I was a dog lover and found no difficulty in
developing a good rapport with Ted. We enjoyed long walks together and I
tried to teach him to walk to heel without the aid of a lead, but Ted quickly
reverted to his old ways.

Ethel had step-brothers and sisters but only one full brother, Thomas. Her
mother died when she was only five and her father married his wife's
spinster sister. Tom had been a master printer and in this sense was probably
the most qualified member of the Bunn clan. In his time the printing unions
were very much a 'closed shop' and master printers earned good wages,
which were the envy of many other trades. With the replacement of 'hot
lead' by the computer terminal, all this has greatly changed, but Tom lived
out his days in relative affluence and security as a respected senior member of
his craft. Tom and his wife Daisy had two children, Joan and Edna, who, with
Bunny's other cousin, Joyce, were bridesmaids at our wedding. Both duly
married and produced seven children between them and this other branch of

Bunny's family also attended the Farnham 'Bunn fights'. As a consequence, we often had as many as thirty or more people enjoying Indian curries and strawberries and cream on our large lawn. Our move to Edinburgh has, naturally, reduced the number of meetings with the extended family. 400 miles is a long way to travel for a Sunday lunch, but we still keep in close touch. Meanwhile our immediate family has grown, with three marriages and five grandchildren – but more of that in chapter 19. I hope this account of the 'two clans' from which our branch of the tree has sprung gives a little flavour of our 'Happy Family', of which we are both very fond.

CHAPTER 18

Friends and Mentors

'A friend may well be reckoned the masterpiece of nature'
Ralph Waldo Emerson

I HAVE BEEN FORTUNATE in having found, or made, very many dear friends during the course of my life; I have also had the good fortune to have known, and worked with, many close colleagues. Many other folk have come and gone at different stages of my career, but a core of true and lasting friendships has been of great strength to me through good times and bad.

I have come to appreciate that all these lasting relationships have contributed greatly to my journey through life in various ways. Some have taught me important truths; others have steered me clear of rocky waters into safe havens. To all of them I owe a great debt and it is impossible to list them all in this book. In this chapter I have endeavoured to sketch some pen-portraits of a few of my closest friends and colleagues, who have had the greatest influence on my life.

My Father, Llewellyn William Charles Montgomery Wilson F.C.I.S.

My father, or Dad as I always called him, was forty years of age when I was born and therefore in his fifties when I was in my teens. Forty years is a big age-gap, hence my memories of my father were of a mature man rather than a youthful Dad. I remember him from my earliest days and can recall painting a little welcoming picture which I pinned to the front window of our Beckenham home when he came back from his London office to a late lunch on Saturday.

Dad was strict but kindly and the most upright man I have ever met – in both senses of that word. He had a back as straight as a ram-rod right up to his 90's and his moral integrity was so deeply rooted that I am quite sure he never suffered from the temptation to cut a legal or ethical corner. At times this could rebound badly on the rest of the family. On one occasion, whilst we were all looking forward to our annual holiday at Frinton-on-Sea, an unexpectedly large gas bill arrived. It was due to the fact that the spring had been colder than usual for an exceptionally long period. Although Dad had a bank account and his creditworthiness was unquestioned, his reaction on

57. My father at work in his London office, Bermondsey, 1949.

receiving the gas bill was to pay it and cancel the holiday! Many would say
he got his priorities sadly wrong and indeed the rest of the family were most
put out by this decision, but no one disputed his motivation.

Dad had a most unhappy childhood and he was scarred by boyhood
memories throughout his life. His father became a widower when Dad was
only four and he subsequently remarried. His second wife was a very strict,
authoritarian woman who showed no affection towards her stepchildren.
They were made to carry out menial tasks before and after school each day,
with no praise or encouragement and with much harshness if there was any
fault to be found. On one occasion my father was instructed to polish all the
family silver, a task which took several hours. On examining the results, his
stepmother came across one fork which still showed signs of the cleaning
fluid between the prongs. The result was that all the silver, without further
ado, was plunged into a bowl of dirty water and the whole lot had to be
cleaned and dried again. When children arrived from the second marriage,
these immediately became their mother's favourites and my father and his
sister were moved from their own rooms to what would have been the
servants' quarters. Dad and his sister left home as soon as they had completed
their schooling. Neither ever re-visited their home or saw their stepmother
on any future occasion. Contact with the family was not entirely severed,

since my Dad recalls meeting his own father on one occasion in the middle of a cold winter and being given a warm overcoat, which he could not have afforded to buy for himself.

My father's first job was that of a junior clerk in the Board of Education. His duties entailed running from one part of the office to another with 'PQs' (Parliamentary Questions) in a file tied up with red tape, waiting for a senior official to draft an immediate reply. This document then had to be taken to more senior civil servants for approval, before the final version was rushed to the Minister's private office. When Parliament was in recess, there were no PQs to cart around and my father's duties then consisted mainly of making tea for his superiors and helping with the filing of the departmental correspondence.

My father was an intelligent person and he decided that the life of a junior civil servant was not what he wanted, so he cast around for more congenial employment. He eventually found a job as an accounts' clerk with the company Charles Southwell and Sons, makers of high class jams, sweet-meats and confectionery. He worked hard and obviously impressed his employers, who sent him to night school to train, firstly as an accountant and later as a Chartered Secretary. With these qualifications behind him he rose through the ranks, ending his career as Company Secretary and Director, in charge of the company's finances.

My father took great pride in his three children and did everything possible to ensure that we got a good start in life. My elder twin-brothers were born at the beginning of the First World War and my mother had a hard life whilst my father was serving in the Royal Navy. I was born in 1928 and although this was during the Depression, I had a better innings than my two brothers. I was sent to a day Public School whilst my brothers went to Beckenham Grammar School. As we have seen (chapter 2), I went from school to university but my two brothers both went out to work on reaching the age of 17. Soon they were caught up in the Second World War – Arthur as a policeman in Palestine and Donald in the Royal Air Force. Perhaps unfortunately, my father showed his pride in my university achievements a little too visibly and this was resented a little by my two brothers, who thought, understandably, that I had enjoyed an easier ride than they had.

Throughout my life my father was helpful and supportive, always giving good advice but never wavering from the strict moral code which was part and parcel of his life. When we first went to Uganda, a car was essential but the University had no scheme to assist staff by providing pool cars or granting car loans. I wrote to my father asking if he could help with the down-payment for a car (I needed about £100), but back came the predictable, Micawber-like reply, 'If you can't afford a car, don't borrow in order to buy one!' In these days my father would probably be regarded as very Victorian, and out of touch with modern-day values, but to me he was a bulwark of moral integrity and this provided me with a standard by which to judge the propriety of my own actions. Although I would not dream of cancelling a family holiday in order to pay a gas bill, I am grateful to him for the good example he set me from my early days and I have tried to live up to these standards. However, I am well aware of the fact that I have not been as consistent in my actions as my father, from his cradle to his grave.

My Parish Priest, The Rev. Cyril Waynforth, A.K.C.

Cyril Waynforth was the Vicar of my local church, St. John the Divine, Selsdon, Surrey. (We lived in Sanderstead parish, which was also our postal address, but we went to church in nearby Selsdon). In fact, he was in reality 'Priest in Charge' for most of his time there, as the church was a daughter church of the real 'parish church' located in Croydon.

Cyril was a saintly man, very sincere in his deeply held beliefs and unshakeable in his loyalty to what he regarded as the 'eternal truths' of the Church's teaching. He was of medium stature, and outwardly had a somewhat stern, perhaps even aesthetic appearance. I have seen him smile but never laugh out loud and I certainly never recall him making a joke. To many, he was lacking in humour and had a somewhat solemn personality. I

58. The Rev. Cyril Waynforth, his wife, Shirley, and his family, Sanderstead, 1948.

would have been only four years old when I first met Cyril on moving from Beckenham to Sanderstead. My parents took me to church three times every Sunday: Holy Communion at 8, Matins at 11 and Evensong at 6.30. In addition, I attended various other functions two or three times a month, so over the years my family and I got to know Cyril very well, although in those days he would have been addressed as 'Mr Waynforth' or, by the more high-church members of the congregation, as 'Father Waynforth'.

Cyril had a curate, Arthur Stevens, who was different from Cyril in every way. Arthur was a bouncy, fun-loving person and a wide smile was almost always in evidence. He tried to put a lighter touch on heavy matters of ecclesiastical teaching, and this meant that on first acquaintance Arthur came across as more approachable. However, in spite of this, it was Cyril that we turned to for guidance and for comfort when tragedy struck, and it was Cyril we held in high regard as a 'Man of God' and whose deep faith we all respected.

When I was eleven years old, Arthur Stevens formed a Youth Fellowship (YF) in the parish and we had weekly meetings in the rather bleak and awesome 'Bell Tower Room' on the first storey of the square bell-tower. Incredible as it may seem today, these meetings were very formal and serious in nature. We studied such academic texts as *The Life and Teaching of the Church of England*, and each meeting started and ended with a prayer, always led by Cyril Waynforth. As our friendship within the YF grew, we organised other less serious activities. We went on Youth Hostelling expeditions and regular cycle rides. We played games, organised dances and ran social evenings and concerts for our parents and friends. All these activities were approved by Cyril, who invariably presided over all meetings of the YF Committee. He kept a firm, but benevolent, finger on the pulse and quietly but firmly let us know if we were stepping outside the well defined mark of a 'recognised' parish organisation.

Looking back, one might have expected the young folk involved to have rebelled at such an authoritative regime. However, I suppose Truman's maxim, 'If you can't stand the heat, get out of the kitchen!' applied, and doubtless we lost some members who found Cyril's way of doing things a little stifling. Nevertheless the vast majority stayed the course and came to love and respect Cyril and to strive to live up to the high standards he clearly laid out for the adolescents of his flock. Before each General Election he would exhort his parishioners to vote, but he would never give the slightest hint as to which party they should vote for. In his book, politics and religion did not mix! On one occasion an elderly Tory supporter put up a notice on the church notice board advertising a coffee morning to raise funds for the local Conservatives. Mr Waynforth took it down and I was present in the porch when he gave the parishioner a sound ticking off for her action. Unperturbed, she replied, 'Don't worry, Mr Waynforth. I know you do not take part in politics. But everyone in this parish votes Conservative, so what's the harm in my little notice?'

Because of his high standing, the majority of the young folk of the parish complied and fell in line with the strict moral teaching which Cyril offered. Sex before marriage, or any unnatural sexual deviation, was abhorrent and strongly condemned. Once safely married, divorce was unthinkable and re-marriage of divorcees completely beyond the pale. These matters were openly debated within the YF, but Cyril's strict moral lead was never questioned. Although healthy youthful flirtations abounded, these early sexual manifestations were always conducted within the constraints of the proper behaviour which Cyril expected from us.

Later on, Cyril was the person much in demand to celebrate our weddings and, if we were still in the area at that time, to officiate at the baptisms of our children. In our case, Cyril had moved to his next parish in Whitstable, Kent. I was very keen that he should be asked to preside at our wedding, which was to take place back in Selsdon church, since my wife Bunny had no especially close affiliation to any particular church at that time. We duly went to Whitstable to receive 'instruction' from Cyril about our wedding, which was simple, straightforward and predictable. Marriage was for life. Children were to be welcomed and should be brought up in the Christian faith. Marriage was between two equal partners in the eyes of God, but where it came to making a difficult decision in such matters as where to live, what job to do, which school to choose for one's children, then, in the case of disagreement, the husband should exercise his right to make the final decision. For this reason, wives were expected to 'love, honour and *obey* their husbands'. The rules were clearly writ; it was expected of us that we would follow them. Strange as it may seem in today's terms, these premises did not give us any trouble or cause disagreement between us. How different to the situation today, where most couples live together before marriage, where divorce is commonplace and any thought that the husband should at any time exercise his 'casting vote' sounds like a case of gross sexual dis-crimination!

Today vicars like Cyril are rare, if indeed they still exist. Current thinking is that we should be more tolerant and not stick too closely to outmoded standards of behaviour. When I was a boy, practising homosexuality was a punishable crime; now the debate is whether homosexual relationships can be sanctified in church as a form of marriage. It is amazing how much has changed in 60 or so years. I am sure that a more questioning attitude to religious and moral issues is a good thing and is to be applauded, but I still consider that I was lucky to be raised at a time when the boundary between 'right and wrong' was much more clearly writ for all to see.

My Friend from School Days, John Stuart Sclanders, C.A.

I first met John when we were members together of the Youth Fellowship run by our local church. We both took our membership of this Wartime Youth Club very seriously, attending most of its activities and sharing the various offices between us. For six years I was the Secretary whilst John, who always showed an aptitude for accounts, was the Treasurer.

By a strange twist of fate, John's father (who was a stockbroker) had planned a University career for John whilst my father (a company secretary) had hoped that I would become a chartered accountant. We were to disappoint both our parents as John, after a two-year stint in the Royal Army Pay Corps, went on to take his accountancy examinations, whilst I proceeded directly from school to University. I have often wondered how our careers would have turned out had our fathers' plans for us come to fruition. Our paths at school went in different directions. I was one year senior to John and was in a different House. Apart from this, I joined the Officers' Training Corps while John joined the Scouts. I was on the Science side of school life whilst John was on the Arts, so our friendship during school days was very much a friendship in spite of our different interests rather than because of them. John found his wife-to-be, Jean, in the Youth Fellowship, and although I made many girl friends through this route, I did not meet my future wife until I was in my last year at Wye College. An interesting anecdote is that on the last of my College Christmas vacations I made up a foursome at a school dance with John and Jean, taking as my partner a former girl friend from Sanderstead. Bunny was not available since she was working and could not get time off. During the evening, as I was dancing with Jean, I told her that I had met the girl I was about to marry – and it wasn't my partner at the school dance! Jean and I have chuckled over this incident many times since, Jean sometimes asking whether she was the first to know of my future intentions and whether I had still to pop the question to Bunny.

In fact the proposal was made formally the following March in a rowing boat on the lake in Regents Park, London, and we were married that September, shortly after which Bunny and I moved up to Scotland. Towards the end of our year in Edinburgh, we invited John and Jean to travel up by train and join us for a week in the Highlands, touring by hired car. John and

Jean were not married at that time, although they were engaged. As fate had it, we each missed the other's marriages because of the unfortunate timing of our respective weddings. The Highland holiday was a great success and took place when it was unnecessary to book in advance, even in the summer months, as the supply of overnight accommodation was greatly in excess of demand. One evening when we were driving south down Loch Ness, we realised we had left the search for accommodation a little late. The usual abundant supply of suitable rooms was unfortunately very scarce. We decided that whatever happened, we would stay at the next place that could accommodate us. It turned out to be a delightful cottage with a B and B sign outside, run by a charming Scottish lady who showed us her only two bedrooms with a double bed in each. Today this would not have posed a problem, but in 1950 things were different and in accepting the rooms we decided that John and I would share one room whilst Jean and Bunny slept in the other. Our hostess was very apologetic, especially when she realised that Bunny and I were newly married, but consoled herself by exclaiming, 'Och weel, the next time ye come this way, ye'll both be sleeping in your proper beds!'

We kept in close touch with John and Jean throughout our 14 years overseas, visiting them on every leave. We were godparents to each others' children and followed their progress through school. They kept a kindly eye on our children when they were at boarding school and we were working abroad. When we returned home in 1964, we saw each other more frequently, attending our children's' respective marriages and following the progress of the next generation with great interest.

In the early 1970's John informed me that he was Secretary to two charitable Trusts, both endowed by Frank Parkinson, a successful electrical engineer from Guiseley, Yorkshire. One Trust, the Yorkshire Trust, ran some very successful Old Peoples' Homes in Guiseley. The other Trust, the Agricultural Trust, was set up for the good of British agriculture. The Yorkshire Trust was running very successfully but the Agricultural Trust, by contrast, was in trouble, mainly because there was no one on the Board who knew anything about agriculture! John suggested that with my background, I might make a useful Trustee and I was duly interviewed and accepted, as already described in chapter 16. For twenty years or so John and I were back in harness together and we have spent many happy days conducting the business of these Trusts. Although we have both now retired from office, having passed the magic age of 70, we are still active Trustees, albeit without the former responsibilities. John has been the best Trust Secretary I have ever met. Meticulous in his control of finances and budgets, but actively involved and interested in the real work of the Trusts. (I recently published a short History of these two Trusts (183).

My Biology Master, Dr Cecil T. Prime, M.A., F.I.Biol, F.L.S.

Cecil Prime, known to his Biology students as 'Cheese' (prime cheese!) was an extraordinary man. In appearance and demeanour he was unimpressive, perhaps a little dour and lacking any obvious sense of humour. To the junior pupils, who only studied biology up to the General School Certificate stage, he was a pretty ordinary, run-of-the mill master, probably soon forgotten when they moved out of his influence to study other subjects. But to those of us who stayed in the science stream and took biology up to the Higher School Certificate level, he was a tower of strength and remembered with great affection throughout the rest of our lives.

Cecil was an extremely productive scientist, with skills and talents known only to the most senior boys. He could easily have obtained a First in the Natural Science Tripos when at Cambridge, but had to be content with an aegrotat degree because he contracted the debilitating illness, Crohn's Disease, which plagued him throughout his life and prevented him taking his finals. He was a Fellow of the Linnean Society and served as its Vice President for a time. He was a founder Fellow of the Institute of Biology and a member of its Fellowship Committee for several years. He was a member of the British Ecological Society and the Botanical Society of the British Isles as well as taking a leading role in local Natural History Societies in Croydon and Surrey. In addition to these professional contributions, he was a leader of the school Scout Troop for many years and a very competent oboist.

Cecil was a pioneer in bringing ecological teaching into the school curriculum. Every Wednesday afternoon he organised field expeditions for such boys as were sufficiently interested, visiting many interesting habitats within an hour's journey of Croydon. He was also one of the small dedicated band of ecologists who set up the *Council for the Promotion of Field Studies*, one of the first centres being based at Mickleham on the North Downs, some fifteen miles distant from the school. In 1970 this led him to publish *Investigations in Woodland Ecology* (with Heinemann) and also his best-known work, *Lords and Ladies*, produced by Collins in their pioneering New Naturalist Series.

Cecil expected his students to work hard and become masters of the botanical arts and many did. He was unstinting in his help for those prepared to give up time to the study of Botany, but somewhat dismissive of, and scathing towards, those who found that the Wednesday afternoon ecology expeditions clashed with the demands of Cricket and Rugby football, or in my case, the Officers' Training Corps. For those that managed to reconcile these conflicting demands on their time, he became a lifelong friend and companion and he took great pride in following the careers of his former pupils. My friend Norman Simmonds (a distinguished plant breeder who was an Old Boy of Whitgift and also a colleague of mine in Trinidad) recently described this unique contribution of an outstandingly brilliant schoolmaster and identified no less than eighteen of his former students who had achieved international recognition in biology. But Cecil Prime not only taught his subject exceedingly well, he also imparted a love of learning and the skills of experimental science. Few schoolmasters could claim as much, and none could have made a comparable contribution to secondary school education in the difficult days of the Second World War.

My Agricultural Mentor – Mr Ken Smith, Farm Labourer

In chapter 3, I described how several of my school summer holidays were spent working on a farm in Kent, where my brother Donald was billeted during the War. During these working holidays, I shared a room with a single farm worker, Ken Smith, who took me under his wing and taught me a lot of what I know about the more basic practical aspects of farming. We worked together on a small-holding belonging to a farmer, George Self, who owned thirty acres of undulating heavy land on the dip-slope of the North Downs. Ken was not raised on a farm and was completely self-taught. He was a small, wiry person, probably in his late 30s or early 40s when I first met him. He had a large bushy moustache and a crop of dark unkempt hair. Although slightly built, he was as strong as a horse and he worked a 16-hour day seven days a week, for very little money and without complaint.

We both stayed in the house belonging to George's widowed mother. She had a small market garden, extending to about 3 acres, carrying an extensive range of vegetable crops, a small orchard shared with a small flock of geese, some rabbit hutches, a duck pond, six chicken runs and a small outhouse converted into a piggery for three sows and their litters. We got up at five and worked in this market garden for two hours before breakfast. We then cycled to George's small farm two miles away and worked in the harvest fields until about six, taking a picnic lunch with us which we ate at midday. At six we returned home to do another two or three hours work in the market garden before enjoying a bath and a very late supper. On Sundays, while I rested and played croquet with my brother Donald and his RAF colleagues, Ken worked as a gardener on a large estate with five acres of grounds three miles away. I found these long hours very difficult for a boy aged between 13 and 17 years, but it was made clear that I was not to work as a gardener on Sundays and that I should only work on Mrs Self's small holding as and when I wished – it was not part of my 'contract' with George Smith. My wages were five shillings a week, plus my keep!

However, Ken knew so much about so many different things that I was anxious to become his willing pupil and if this meant working extra hours, then so be it! Every day there were eggs to be collected, broody hens to be caught and put into coops, mature chickens to be killed, de-feathered and dressed, rabbits to be skinned, drawn and quartered, vegetables to be gathered and prepared for market, fruit to be picked, land to be dug over, compost to be made and applied, animals to be fed and watered, grass to be mown, hedges to be cut and weeds to be tackled with scythe and sickle. But this was only the voluntary early morning and evening work. By day there was hay to be mown, turned by hand, carted and stacked, corn to be cut with reaper-and-binder, sheaves to be stooked, stooks to be broken down and carted to the stackyard, corn stacks to be built and thatched with corn dollies firmly attached. All these tasks required different skills and the right implements with which to perform them, varying from the intricacies of the binder mechanism in the reaper-and-binder to starting up (by crank handle) and driving an old Case tractor. I learnt to drive this tractor and to master the use of its assortment of specific tractor-drawn implements, each with their own individualities and functional peculiarities. Ken was master of them all, and I was fascinated to learn as many of the skills as my brain could accommodate, so that I didn't need to have the same operation explained to me more than two or three times at most.

Ken was considerate, gently swearing at me when I got things wrong, but warmly congratulating me when I got them right. He was at all times a real friend and companion, and he seemed to enjoy teaching me. We only violently disagreed on one major matter. I told him that I was intending to proceed from school to Wye College to study agriculture and this news was greeted with scorn and contempt by Ken. 'Why go to College?' he enquired, them b—s don't know nothing about real farming – it's all book learning! There's nothing you'll ever learn from them you can't learn from me! Most of the b—s at college have never done a decent day's work on a farm in their life!'. In many ways, of course, Ken was absolutely right. I never learnt to skin a rabbit, pluck and dress a chicken, drive a tractor, repair the binder mechanism at College, although I did learn a lot about what made rabbits, chickens, tractors and binder mechanisms tick. But perhaps the biggest tragedy was that Ken firmly believed that if you mastered all these practical skills and worked conscientiously all the hours of the day, then farming would always be profitable. He did not appreciate how quickly things were changing and how it would not be long before three-acre small holdings and thirty-acre farms would no longer be viable, so that many good honest country folk like George Self would go out of business.

Men like Ken Smith are the salt of the earth. For many centuries farming has relied on the skills and hard physical labour of people of his ilk. He was one of the last of a generation of good all-round country folk and it is unlikely that when they have left this earth, they will ever be replaced.

My College Contemporary, Mr Stephen J. Carr, OBE, BSc.

Stephen was a contemporary of mine at Wye College. We were not in the same year, as I managed to get direct entry into the second year of the course whilst Stephen had to start, more logically, in year one. We were both members of the Student Christian Movement (SCM) and amongst the jobs we shared was that of taking services at the small mission church at Knackholt, a village south of Wye. We were both also friendly with a family, the Butchers, who lived in Brook, a nearby village. Eric Butcher was an ex-Palestine policeman, which was of interest to me as my brother Arthur had served in the Palestine Police. Iris Butcher had been a missionary in Palestine and they were married there just after the War. They had two daughters, Patience and Angela, and Stephen and I (and our then respective girl friends, Bunny and Anne) used to baby-sit for the Butchers when they went out. Although I did not know it at the time, Stephen had always been interested in becoming a missionary and whereas my point of interest with the Butchers was with Eric and the Palestine Police, Stephen's was with Iris and the missionary field.

After College Stephen and I went our respective ways. I went to Edinburgh and thence to Africa, whilst Stephen eventually became a missionary in the South Sudan. His missionary activities were invariably connected with agriculture. He could see no purpose in converting people to Christianity unless he could also offer them improvements to their daily lives and since he always worked in agricultural communities, this meant assisting them to farm better and to market their produce more efficiently. Wherever Stephen and Anne went, they always left behind oases of improved agricultural practice centred on a mission church (and often a mission hospital). Faith and better standards of living were, in Stephen's mind, intrinsically linked.

We next came in contact with Stephen when we were working at Serere, in North-east Uganda. Stephen was then in Torit in the Southern Sudan, which at that time was undergoing civil war. The Christian south had always been under threat of attack, or actual attack, by the Arab north and in the late 1950s the hostilities had reached a high pitch and it was clear to Stephen that he and Anne had to leave if they were to survive. He had picked up a debilitating tropical disease and was grossly malnourished when he eventually crossed safely over the Sudanese/Ugandan border. Somehow we heard of his arrival and immediately invited them to stay with us in Serere, which they did. He was close to starvation when he arrived and extremely ill, but rest and good food worked wonders and it was not long before Stephen was restless to be off and find another outlet for his missionary activities.

Stephen worked in numerous African countries and for a time he was employed by the World Bank as an expert on rural affairs. I did not have much time for many of the so-called experts from the World Bank. Many justified the tag, 'A man with a briefcase five thousand miles from home', but Stephen was probably the best and most experienced person they ever employed. He undertook agricultural extension not by precept but by example. He never advised small farmers to do anything he could not do himself. He learnt the local languages wherever he went and many good African folk will have cause to remember Stephen and Anne Carr for a very long time to come.

Stephen and I have both kept in touch, not so much as a consequence of our work in Africa, but through our life-long support for Wye College and its old students' association, The Agricola Club. Stephen became a representative College Governor and was subsequently elected a Fellow of the College. When I joined Stephen in the Fellows stalls a few years later, I was immensely proud to be greeted by him, as I regard Stephen as a truly living example of unselfish living, with a life truly dedicated both to God and to the betterment of the human race.

My Postgraduate Examiner and Reader in Animal Physiology at Cambridge, Sir John Hammond, C.B.E., F.R.S.

Plate 59 shows John Hammond as I remember him. He was very tall, usually most informally dressed, and whatever the weather, he wore an old shabby raincoat. He sported a somewhat unkempt multi-coloured moustache, burnt by the ever-present drooping cigarette. He rode an old bicycle round Cambridge even when going to an official dinner party. John Hammond never learnt to drive a car and it is said that his bicycle was so old and decrepit that there was never any risk of it being stolen.

I first met John Hammond when he visited Newcastle to deliver a lecture on animal science to the UK Agricultural Students' Society. His lecture made a great impression on me. Hammond was clearly a master of his subject and

59. Sir John Hammond, FRS, Cambridge, 1959.

one who had been a major influence on my postgraduate supervisor, Mac Cooper (see chapter 5). He had made notable contributions to most branches of the subject and had published several text-books on subjects ranging as widely as animal breeding, animal nutrition, animal physiology and systems of animal production.

As I have explained, I could not graduate until I had been in residence at London University for three years. Therefore I became a 'premature postgraduate' and, under the supervision of Mac Cooper, worked on the *Effect of Plane of Nutrition on Growth and Development of Gallus domesticus.* When the time came for my M Sc thesis to be examined, John Hammond was the choice as external examiner. The examination took three hours, but the time flew past. John Hammond went through my thesis page by page, giving helpful comments. And so it was that, some months later, three papers were sent to John Hammond, then editor of *The Journal of Agricultural Science*, for approval prior to publication (1,2,3).

Whilst abroad, I continued to work in fields pioneered by John Hammond and he always kindly took the same care with my typed manuscripts as he had done with my M.Sc. thesis. Some of these papers were critical of John's earlier work, in that they suggested modifications to his experimental design and at times came to somewhat different conclusions. Far from resenting this, John was delighted that the subject was 'being pushed forward', and he seemed genuinely pleased that I was following in his footsteps. During the course of my home leave I always spent some time with John Hammond at Cambridge, talking about research matters and thanking him for his help. It was difficult in Africa to have access to a library to look up research references which were demanded by journal editors. Some required first and last page numbers for each reference, others needed the correct initials for all quoted authors. If any of these data were missing, Hammond would do the necessary library research on my behalf, to ensure that everything was in good order for publication.

On one occasion I blotted my copybook with John Hammond. He was leading a British delegation to a meeting of the European Association of Animal Production in Seville, Spain. We flew to Madrid, where we had to transfer to a local flight, but all the announcements were in Spanish and, as I was working in Trinidad at the time, I had acquired a smattering of that language. Over the tannoy I heard part of an announcement which I took to be, '. . . Seville . . . Puerto Quatro . . .' I gathered all John's party together and bundled them to door no. 4. There was a bus waiting which we all boarded, although there were very few other passengers with us. However, the driver duly drove off, through the main entrance to the airport leading back to the city of Madrid. In panic, I went up to the driver asking him whether he was taking us to the plane for Seville. 'Oh no', he replied in Spanish, 'I am taking you to the Hotel Seville in Madrid!' It took a great deal of persuasion to get him to turn round and head back to the airport, where thankfully our plane was still awaiting us. John Hammond turned to me and remarked, 'Perhaps they speak a somewhat different sort of Spanish in the West Indies!' My companions did not let me forget this mistake for a very long time.

When the time came for John to retire from his Readership in Animal Physiology at Cambridge, there was considerable alarm. While some aspects of his work were continued, much of the research programme was stopped. However, Hammond was still active in retirement. He was retained in an advisory capacity by many organisations and he continued to travel the world giving advice on a wide range of topics. His extended family of ex-students, whether Cambridge folk or not, are forever in his debt. Recently I was delighted to be invited to give a paper in Cambridge about John Hammond and his work on the occasion of the opening of a new Equine Research Centre. It was good to see many of John's previous co-workers and also to meet his son, John Hammond junior, and his family.

I was even more privileged when I was invited by the Council of the British Society of Animal Science to deliver the 50th Anniversary Hammond Memorial Lecture, in 1994. I described John Hammond's many achievements, one of which was helping to create the British Society for Animal Production (as it was first called). I then went on to consider the changes which the Society might have to consider making over the next 50 years (178). As a result of this Lecture, I was invited by the Council to chair a Working Party to propose changes to the Society's structure and working arrangements, to enable it to become more proactive and less reactive in responding to change. I regarded this work as my own private tribute to John Hammond and a small token of gratitude for all that he had done to help shape my own career during the previous 50 years.

My Colleague in Trinidad, Professor Willem Boschoff, MSc., Ph.D.

Willem Boschoff was a South African, but a South African with a big difference. Although of Dutch descent, and brought up under the Apartheid system at Standerton, there was not a trace of racist feeling in his character. After taking a postgraduate course in agricultural engineering at Newcastle University, Willem was recruited by Makerere College as one of the first new appointments. These followed upon the arrival of Professor Fergus Wilson as Head of the Agricultural Faculty, of which, until 1951, I was the only member. Willem was stationed on the main University campus in Kampala, whilst I was working two hundred miles away at Serere. Thus we met only when I came down for staff meetings in Kampala or when Willem came to visit me up country.

We quickly struck up a friendship. We had the same passion for working for the good of African agriculture and we shared the same sense of fun. I have already recounted how this humorous streak was not appreciated by Fergus Wilson, who was fearful that we were pulling his leg behind his back, so to speak, so that he never knew when to take us seriously. Willem was more expert at poking fun at people and situations than I was, but I endeavoured to emulate his enthusiasm for a joke and this often landed both of us in trouble.

Willem embarked upon a programme of work for his doctorate at the same time as I was working on my growth project with the East African dwarf goat. He chose as his subject the production of methane gas from vegetable waste. Fergus had always interested himself in compost-making and had published papers on the subject when he was working in Zanzibar, so Willem deemed it expedient to ask Fergus to become his Ph.D. supervisor. Indeed, there was little choice, since there were no other engineers on the University staff at that time. Willem thought it prudent to bring Fergus into the frame by asking for help in the matter of the temperatures generated in heaps of decomposing vegetable matter, which affected the speed of production of methane. 'I don't actually know how high the temperature rises,' Fergus admitted, 'but I do know that it gets very, very hot when you put your hand into the material'. Willem subsequently devised what he called the Fergus Wilson temperature scale on this basis. This had three grades of hotness: 'Warm'; 'Fairly hot'; and 'Great Scot – bloody hell!'

Willem had an excellent rapport with his students. He was a good lecturer and his time was always at the disposal of his class. He was very concerned at the barriers which divided expatriate officers and senior African staff and did all he could to break these down. The barriers extended into the sports fields and Willem, who was a first-rate rugby player, did all he could to bring African players into the game, including Idi Amin. On one occasion, when Uganda were playing an away game against West Kenya, the host committee demurred about the African players in the Uganda XV, saying that their local 'supporters would not like it'. This incensed Willem, who promptly said if that was the reason, the match would have to be moved to Uganda, where his supporters would pose no problem.

After serving at Makerere for twenty years, Willem moved to West Africa and later to Malawi. During this period Willem penned his own auto-biography (176) and kindly sent me a copy. It was this initiative of Willem's that made me more receptive to the suggestion that I should also endeavour to set down my own record, which overlapped with that of Willem Boschoff for six interesting years.

On retirement, Willem returned to his native South Africa, where he has a small citrus farm. He is still furthering the cause of landless African farm labourers and has settled many of his former staff on plots of land hived off from his farm. In spite of the change of government, establishing legal tenure of his land for his workers has been fraught with countless bureaucratic difficulties. Willem was a good friend and colleague and I greatly admire his continuing stand for African rights.

My Professor in Trinidad, Professor C.C. Webster, C.M.G., BSc., Ph.D., A.I.C.T.A., F.I.Biol.

Tom Webster, like me, took his first degree at Wye College and in due time we were both elected Fellows of the College. Tom's tropical career took him to Cambridge and thence to Trinidad to take his AICTA Diploma, after which he served in Africa, firstly in Nigeria before the War and then in Kenya during the hostilities. In 1950 he left Africa to become the Deputy Director of Agriculture in Malaya until being posted to the Chair of Agriculture in the Imperial College of Tropical Agriculture (ICTA). I first met Tom in Trinidad, where he had been partly responsible for my appointment to the Senior Lectureship in Animal Production, just before he took up his own appointment in 1957.

I was immediately impressed by Tom's administrative capabilities. He had the ability to bring out the best in people in a gentle way, whilst at the same time he could be highly critical of sloppy work, or failing to attend to the joint needs of students and research. His own lectures were quietly given, in a matter-of-fact manner, but they were greatly appreciated by the postgraduate students, who respected him for his knowledge and experience.

His interests lay on the crop side of tropical agriculture and he was the acknowledged world expert on the tung oil tree, on which some of his early research was based. He was keen to encourage a high standard of research by his staff, and was helpful in finding financial support for their projects. Having had a career in the Colonial Agricultural Service, he was experienced in the ways of the Colonial Office, which were often a closed book to members of staff who had arrived in Trinidad from other Universities or from industry.

Tom had a keen sense of fun and although his own brand of humour was somewhat on the dry side, he invariably had a twinkle in his eye. When meeting new staff or getting to know his students, he adopted a listening pose, saying little but probing by eye-to-eye contact rather than by asking a string of questions. Many folk found this somewhat disconcerting and often they lost a major argument simply by talking themselves beyond their boundary of knowledge. Tom would bring them up short by a raising of his eyebrows, as if to say, 'Do you really mean that?'

In 1959 Tom was invited by Longmans to produce a textbook on Tropical Agriculture and he invited me to become his co-author and write the animal chapters of the book. I accepted, little knowing that one year later Tom was to be invited to take over the Directorship of the Rubber Research Institute in Malaya, after which he transferred to become Director of the Oil Palm Research Institute. It followed that our conjoint bookwriting activities had to be conducted at arm's length. As a consequence, the text was not published until seven years later (39). It was well received and was reprinted several times, so in 1975 Longmans asked us to produce a second edition. This we did, again by distant correspondence, and this new edition also went into several reprintings (110). We were again commissioned to produce a third edition in 1990, but this time things were much more difficult. Tom's eyesight was failing. His ability to do the necessary library work for the third edition was severely impaired, and other writers had to be drafted in to assist with the revision of the crop sections. This time, the gestation period was even longer, and the edition did not come off the presses until 1998 (182).

Tom returned from the Oil Palm Research Institute in 1965 and took up the post of Scientific Adviser to the Agricultural Research Council (ARC). He did this job superbly well and I saw a lot of Tom over the next ten years, since my work brought me into regular contact with the Research Institutes and with the Council responsible for them. Many holders of Tom's post regarded it as an office job, dealing with the Institutes by telephone call and correspondence. Tom disagreed with this approach and spent at least half of his time 'on the road', visiting the various Institutes and getting to know their scientific staff personally. He is still remembered with affection by the staff who were active in the 1965-1975 period. I too, although not a member of the ARC, learnt a lot from Tom by watching his activities from my position in Unilever during that time.

Tom Webster was a modest man, an outstanding tropical agriculturist and a very effective Chief Scientist of the ARC. His C.M.G., which he was awarded in 1966, was richly deserved and I have much to thank Tom for, in terms of his unique experience with developing countries, his effective style of teaching and his outstanding abilities as a research administrator. When Tom left Trinidad to take up his new appointment in Malaya, he actively encouraged me to put in for his Chair and his support must have been responsible for my success. I will always be grateful that our two paths crossed for those three short years in Trinidad.

My Friend in the Civil Service, Sir Emrys Jones, BSc., F.R.Ag.S.

In the many obituaries written about the life of Emrys Jones, the common theme running through them all is that he was probably the most positive influence on farming in Britain. This influence extended from the War years, when food production was of great national importance, up to the time of UK entry into Europe. Emrys was the son of a Welsh hill farmer. He was educated at the University College of Wales, where he obtained a First in agricultural economics, after which he embarked on a life-long career of agricultural advisory work and education.

When I returned to the UK in 1964, Emrys was in the top job in the National Agricultural Advisory Service (NAAS), and in 1967 he became head of the re-launched organisation, the Agricultural Development and Advisory Service (ADAS). He did not run ADAS from his office in London, but was out and about, meeting staff and addressing farmers' meetings. It was on the 'conference circuit' that I first got to know him well. Many government officers were unhappy about co-operating too freely with 'commercial folk', since they wanted to preserve their independence. Emrys disagreed. He saw British Agriculture as a complex network of farmers, advisers, research workers, folk in commerce and civil servants with statutory duties to perform. In his judgement, if a person was doing his job well, he was a fit colleague to work with as part of the overall pattern. He therefore visited research stations, both in and outside the public sector, and was happy to share a conference platform with a representative of industry, providing that person had something useful to say.

When I moved to BOCM Silcock in Liverpool, our public relations officer, John Deeley, was already on good terms with Emrys from the days when John worked for the Ford Motor Company and Emrys was pushing hard for the efficient deployment of farm machinery as a means to increasing output. Through John's efforts, Emrys was a frequent guest speaker on our company platforms, and as Agricultural Director, I was either in the chair at such meetings or giving a paper to complement the one by Emrys. Emrys enjoyed his food and wine and was a great raconteur. Through John Deeley's good efforts we spent many congenial evenings together in different parts of the country.

We only disagreed on one matter and that was the UK entry to the Common Market. My company, Unilever, were committed to supporting Britain's entry, as they thought that this would be essential for the long-term prosperity of the UK, which could not hold out as an independent off-shore island of continental Europe. Emrys was sceptical, since he could see that we would lose control of our Agricultural policy and would be dragged down to the less progressive agricultural practices dominated by large numbers of European peasant farmers. In a sense we were both right. Joining Europe was, in spite of all the frustrations, in our best long-term interests, but the Common Agricultural Policy (CAP), dominated by the large farming lobbies in France, Germany and Italy, was certainly a heavy cross which British farmers have had to bear. There is now general realisation within Europe that the CAP must change, but how soon and in what way are still open questions. Our greatest mistake was not to join the EC from the beginning, when we could have helped formulate the agricultural policies from within.

Growing increasingly frustrated at the 'European brake' on British agriculture, in 1973 Emrys took early retirement and went to the Royal Agricultural College (RAC) as its Principal. We saw even more of each other during his five years at Cirencester, during which time we again shared numerous platforms together. Emrys acted as Chairman of the Judges' Panel for the various awards which our company made to deserving members of the farming community. At the time that Emrys accepted the RAC post, I had recently been made the Agricultural Trustee of the Frank Parkinson Agricultural Trust and he was very quick off the mark in submitting a request for a grant from this Trust to build a new lecture hall and resource management complex. In due course, when the money had been spent, Emrys laid on a grand Opening Ceremony, starting with an excellent dinner at the College. Of the many grants the Trustees have made since then, none has surpassed the meticulous planning of the RAC project. Later on, when I formed the Three Thousand Gallon Club (see Chapter 11), Emrys 'launched' this initiative at Cirencester with another superb dinner, the quality of which has never been surpassed.

Emrys was a master organiser, who taught me a lot about running successful agricultural events. He was also a most outstanding civil servant, who regarded his chief 'master' to be the British farmer. Although loyal to the many Ministers whom he served, he would never give way to political pressure on matters which he thought were not in Britain's best agricultural interest. When this was known and understood, he developed a rapport with Whitehall that was quite unique. It was a 'partnership' agreement, not the 'Yes, Minister' relationship which is all too common in the ranks of the civil service.

My Opposite Number, Professor J.M.M. Cunningham, C.B.E, BSc, Ph.D., FRSE, F.I.Biol., FRAgS.

Ian Cunningham was a farmer's son and rightly proud of the fact. He was raised on a small sheep farm in central Scotland. He took his first degree in agriculture at Edinburgh University and soon after joined the staff, rising to the rank of Senior Lecturer. He was active in research, most of it connected with sheep husbandry and grass production, and it was therefore natural that I should make a pilgrimage to Edinburgh University on my leaves when working in Uganda and Trinidad. Ian was a good communicator and congenial host and on many occasions he kindly invited me to stay at his lovely home in Edinburgh. We kept in touch between my two-yearly visits and exchanged research papers over many years. I was able to make some small contribution to Ian's work, for when I was the Agricultural Director of Silcock and Lever Feeds, I arranged for a grant to be made to Edinburgh University for the erection of a sheep in-wintering house on the Bush estate. At the time, which must have been about 1970, the idea of keeping pregnant ewes in sheltered accommodation prior to lambing was unusual, although not unique. The Edinburgh sheep barn, high up on Castlelaw Hill, attracted quite a lot of attention from Scottish flock masters, as well as providing good facilities for Ian's research.

Ian was eventually promoted to the Directorship of the Hill Farming Research Organisation in Edinburgh and was responsible for designing and building a new head office with attached laboratories on the Bush Estate. Some years later Ian was invited to accept the Chair of Agriculture at Glasgow University, a position coupled with that of Principal of the West of Scotland College of Agriculture in Ayrshire. This soon became another research facility, which I visited as often as I was able during my 19 years with Unilever and our mutual friendship and respect grew during this period.

In 1983 the Chair of Agriculture at Edinburgh University and the linked post of Principal of the East of Scotland College of Agriculture became vacant and several of my friends, including Ian, persuaded me to put my hat in the ring. I got the job. In 1984 I found myself in the interesting position of being both a colleague of Ian, but at the same time one of his two rivals, since the Principals of the East, West and North Colleges were competing for diminishing amounts of finance. Ian fought his corner strongly and he had a major advantage over me as he knew the administrative system inside out, having worked with the Scottish Office for thirty years. I, on the other hand, was new to the game and a little wet behind the ears.

Ian could not have played his cards with more fairness. He taught me the ropes, whilst never letting me have an easy time. When eventually the time came and the three Colleges had to work more closely in unison, Ian always took the broad view of each situation as it arose and never played tricks with the unwritten rules of the political game. I learnt a lot about the need to fight hard but fight fairly in our dealings with our paymasters and Ian and I never fell out, even though the stakes we played for were high and our respective staffs expected us to ride roughshod over the competition.

During this time Ian and I saw each other at least weekly and communicated with each other almost daily, but our friendship remained steadfast. When eventually first Ian and then I retired from our respective posts, we slipped back into our former habits of enjoying close family friendships, in which Bunny and Ian's wife, Nancy, played an important part In retirement, Ian and I served on many committees together, many of them dealing with agricultural research and the selection of suitable candidates for research prizes. We both served on the Business Committee of the General Council of Edinburgh University, Ian becoming a member of the University Court. Ian became Chairman of the Scottish National Trust and of the Board of the Macaulay Research Institute, whilst I served as a member of the Royal Society for the Protection of Birds Scottish Committee and as General Secretary of the Royal Society of Edinburgh. Thus we both continued to take an active interest in matters relating to agriculture and the environment.

But in all these matters Ian was my guide and mentor. He knew the 'Scottish system' from top to bottom and whatever contribution I have managed to make to agricultural matters in Scotland is in large measure due to the many years of Ian's friendly co-operation.

My Friend, Colleague and Vice Principal, Mr J.L. Beveridge, M.B.E., BSc., MSc.

I did not know Lawrence Beveridge until I moved to Edinburgh in late 1983. Lawrence was then Assistant to the former Principal, Professor Noel Robertson, whom I replaced in January, 1984. Lawrence was the first member of the College staff to invite me to his home, where I met his lovely wife, Margaret. Lawrence and I were close friends and colleagues for the next fifteen years and it was a very sad day when I learnt that Lawrence had died suddenly at home, in June 1999, shortly after his own retirement.

Lawrence started life as a West of Scotland lad, having been born in Drumchapel and educated at Glasgow University and later in America. Lawrence started his academic career at Selwyn College, Cambridge as a Demonstrator in Agricultural Economics. He then returned to Scotland, firstly to the West of Scotland College at Auchincruive and then to a lectureship in the Agriculture Department of Edinburgh University, where he was a colleague of Ian Cunningham's for many years.

Lawrence was a great diplomat and excellent 'man manager'. No problem was too unimportant for him to solve and when I took over at Edinburgh, I found a right-hand man who was steadfast, a shrewd judge of people, a meticulous administrator and a staunch and valued friend. I have worked as a member of many teams during my career and I have never had the privilege of working in tandem with anyone as trustworthy and efficient as Lawrence. Often it is difficult for the person in charge of an organisation to delegate responsibility for everything whilst away on business or on holiday. With Lawrence I knew that everything would be conducted with the utmost diligence and integrity and that if anything went wrong, it was something that no one could have avoided.

A year or two after my arrival in Edinburgh in 1983, Lawrence and Margaret decided to sell their Edinburgh home and move out to a lovely house in the country. Unfortunately difficulties arose in selling their Edinburgh house and for several months Lawrence was the unfortunate owner of two properties, with the consequent financial problem of a large bridging loan. I almost dreaded the recurrent question which I had to ask, 'Any luck with the house sale?' since I knew the answer would probably be negative. In spite of this stress, Lawrence went about his daily work as though he had nothing whatsoever on his mind. He was an example to us all, never letting personal problems get in the way of his work.

My last years at work, as the Head of the School of Agriculture at Edinburgh, were happy and eventful years. The momentous changes were forced upon us by external forces, but the happy environment was in large measure due to the fact that I had Lawrence by my side as friend and colleague. I owe him an immense debt of gratitude, and hope and pray that I communicated my heartfelt thanks to him before his untimely death.

Chairman of the College Trade Union, Mr Jack McGowan, HNC, MRIPHH

Jack was Chairman of the Trade Union Side of the negotiations which took place on a regular basis between 'Management' and 'Staff' of the East of Scotland College of Agriculture during my time as Principal. As such, Jack was in an important position and one which could either have resulted in constant friction between the two sides, or else assisted the organisation to work more smoothly for the common good. Jack was consistently on the side of constructive co-operation, whilst always prepared to take up with vigour any cases in which he felt that staff had a legitimate grievance about management decisions which affected them. When not engaged in his Trade Union duties, Jack was a Senior Technician in the Nutrition Department of the College.

Jack, together with his Union Secretary, held regular monthly meetings with me, my Vice Principal, Lawrence Beveridge, and the College Secretary, Stuart Ainslie. Thanks to the attitude adopted by Jack, these meetings were never confrontational. Real efforts were made by both sides to resolve minor issues before they became major bones of contention.

On one occasion I had reason to dismiss a member of staff who was a staunch member of the Trade Union. He appealed to the Board of Governors against my decision, which was his right. In the course of time he was invited to attend a meeting of the Board to put his case and he exercised his right to be accompanied by a 'friend', in this case Jack McGowan. Jack pleaded for his colleague with great eloquence, listing the good things his colleague had achieved during his service with the College. The Governors supported my action and the appeal was dismissed. Next time I saw Jack, he said, 'Principal, you were entirely correct in the action you took but you realise it was my duty to do the best for my colleague.' 'Jack', I replied, 'I admire the way you put up such an excellent defence. If ever I get myself in trouble with my Board of Governors, I will ask you to represent me as *my* friend!'.

The School of Agriculture Building was a rather unfriendly place, which I thought could be improved by some appropriate floral decorations. The difficulty was finding someone to look after the plants, especially by watering them through the summer months. Jack was a keen gardener so I asked him to take on this job on a voluntary basis as I had no money available. Jack took the job on and performed it in most diligent manner. When Jack went on leave, he asked me to arrange a substitute to tend the plants in his absence. I decided that as I had invited Jack to take on the task in the first place, the least I could do was to offer to do the job in person. Bunny helped me with the task, as she is a better horticulturist than I, and it took us about two hours, twice a week, to complete the assignment. I was therefore concerned lest Jack would not appreciate how much I, and the College, owed to him for his contribution. I decided to give him and his wife a couple of days each year attending one of the national Gardening Exhibitions and to pay for his expenses. This we did and Jack was overjoyed at this small gesture, which was little enough compensation for his many hours of voluntary effort.

On another occasion, at one of our regular monthly meetings, Jack informed me that the staff had voted for a 'No Smoking' rule in the public rooms of the College. I was very amenable to making most of the College a 'non-smoking' area, but I doubted my ability to get away with it, knowing that the move would be bitterly opposed by some members of staff. I put this point to Jack. 'Don't worry, Principal,' he said. 'You make it clear that you are bringing in the non-smoking rule as a result of representations received by my Union. We will undertake to see our members put the ban into effect!' And so it was; I duly decreed most of the College a non-smoking zone. I never heard a word of complaint but I have no idea how the Union managed to keep their end of the bargain so extremely well.

People like Jack McGowan are the salt of the earth and much of the smooth running of large organisations is dependent on their good sense and co-operative spirit. I learned a lot from Jack and have always admired the sensitive way in which he dealt with highly contentious issues. The praises of such folk are only rarely sung, and I am pleased to give this acknowledgement not only to Jack, but to other good folk like him.

My wife – Mrs Bunny Wilson, BSc.

The person who has been of the greatest support to me throughout my career has been my wife. She has been a lover, mother, friend, companion and colleague – and at times my most respected and acute critic. It would be presumptuous of me to include her in this chapter amongst those categorised as 'friends and mentors' and she deserves a chapter to herself. Thus Bunny becomes the major player in the final chapter of this book.

CHAPTER 19

Married Bliss

'But to see her was to love her,
Love but her, and love for ever'
Robert Burns – Ae fond kiss

I DESCRIBED IN CHAPTER 5 how I first met Bunny in the Wye Hill Cafe whilst reading Churchill's *War Memoirs*. It was 'love at first sight', although it doubtless took me quite a time to raise my eyes above the pages of the book. I worked on the theory of 'safety in numbers' in those days. I had about four girl friends with whom I kept up close friendships. One was in Sweden, another in Northern Ireland, a third living in Sheffield and the last one, Betty, who had been my special girl friend for most of the time I had been in the Youth Fellowship, was living in Selsdon. This complicated arrangement led to endless difficulties and it was not long before I decided that Bunny was the only girl for me and the rest should know that I had finally made up my mind about the girl I would marry. I remember spending a difficult evening writing carefully worded letters to them. I then invited Bunny to post them. One thing I think I can say for myself: I am not averse to taking important decisions!

A week or so after posting the letters, Bunny and I went for a long walk on the North Downs of Kent. The weather had been wet for some time and the going was very damp underfoot. At one low spot on the walk we came across a large puddle – in fact, more of a shallow pond than a mere puddle – and I thought it would be a gallant gesture to carry Bunny the few yards to the other side. As we reached the mid-spot, the mud became very slippery and I stumbled, depositing Bunny fairly and squarely in the muddy water. She got up, wet and bedraggled, laughing outright at the look of horror on my face. She has often reminded me of this incident, never quite knowing whether the tumble was a real accident or if it was done on purpose to find out whether she had a sense of humour! I have assured her many times that the fall was not premeditated; whether or not she has believed me is another matter.

Although there was no doubt for either of us that we would marry each other, at that time there was no formal engagement between us. I wanted to do things the old-fashioned way and therefore waited to ask Bunny's father for permission to marry his daughter. This I duly did on March 13th, 1950. William Bunn took his responsibilities seriously, albeit with a broad smile, asking whether I was in a position to support his daughter in the manner to which she had become accustomed. 'I don't think I can', I replied, 'I rather think she will have to support me in our first year of marriage as I will be just a student!' However, permission was duly given and later that day I took Bunny for a trip in a rowing boat on the lake in St. James' Park, London and formally proposed to her. This time, I took good care to see that I did not tip Bunny accidentally into the water!

The next six months rolled by very quickly. As I had close connections with the church of St. John the Divine at Selsdon, we agreed that the marriage should take place there. Because of my friendship with Cyril Waynforth, we agreed that we would invite him to take the wedding service, with the permission of his successor. Cyril had moved to Whitstable in Kent, so we made our way to his Vicarage where he duly 'prepared' us for the wedding and we rehearsed the details of the service. (I have already referred to this meeting in chapter 18).

Although Bunny's parents had agreed to the wedding being held in my church and not theirs, they made one stipulation: Bunny had to leave her family home in her wedding dress and go from there to Selsdon. The journey across London was a long one and required a detailed knowledge of the road network. As the War had been over only a few years, sign-posting in most of England was still suffering from the removal of all useful road signs in anticipation of a German invasion. I therefore borrowed my father's car and Bunny and I drove over the route several times to make sure she knew the critical intersections. Although she would be coming by hired car, I had more faith in Bunny's navigational skills than those of the chauffeur. We planned the timing carefully, leaving about one half hour after the 'expected' arrival at Selsdon for Bunny and the bridesmaids to freshen up at a friend's house near the church.

At the appointed time I duly arrived at the church from my home, together with my best man, my brother Donald. The wedding was due to start at 11 o'clock, but 11 came and went and I got a little nervous. 'Don't worry,' said Don, 'it's the bride's privilege to arrive a little late and in any case she has had a long way to come.' Another ten minutes went by and once again Don tried to reassure me. 'It's a very wet day,' he said, 'and probably there are more people than usual on the road, so there's nothing for you to get worked up about.' Another ten minutes went by, and I said: 'It's not like Bunny to be late. She really is a very punctual person.' 'Perhaps the stop-over at your friends is taking longer than expected.' he said. 'All the bridesmaids have got to powder their noses and your friend has probably given them a cup of coffee. They'll soon be here.' Another ten minutes went slowly by. 'What on earth do you think has happened?' I enquired of Don. This time, he had run out of soothing things to say. 'I haven't a clue.' he said, 'It looks like a bit of a disaster to me!' This didn't do a lot for my self-composure but within a few minutes of our last *sotto voce* chat, the organist struck up the Bridal March and Bunny and her bridesmaids came sweeping up the aisle looking absolutely radiant. We were very fortunate that there was only one wedding that Saturday, or she might have come to find an entirely different wedding in progress.

The cause of the delay was not want of proper planning. As the bridal procession was wending its way through the centre of Croydon, a cyclist just in front of Bunny's car got his wheel stuck in a tram line and fell off his bicycle. Bunny's leading car stopped, but the next two cars skidded and some minor damage was done. This necessitated a wait until the police arrived and this little accident cost the party about half an hour of valuable time. Not only that, but after the accident Bunny's chauffeur tried to make up some of the lost time and in so doing lost contact with the car behind containing the bridesmaids. Bunny did a quick turn round at my friend's house, but there was no sign of the lost bridesmaids. After waiting ten minutes, Bunny took the decision to go ahead without them. Fortunately the bridesmaids did manage to locate the church and they arrived just as Bunny was about to proceed up the aisle. Wisely our daughter and her husband-to-be were content to marry at our local church 25 years later.

We spent our honeymoon at the small town of Westendorf in the Austrian Tyrol. Our hotel charged us two guineas a week for full board, but the cost of the rail fare to get from London to Austria was about ten pounds return, giving a bargain honeymoon costing less than £25. We have revisited Westendorf once since 1950 and we called in at our hotel for a cream tea. We met the new owner, the son of the proprietor who had looked after us during our honeymoon, and he found the 1950 hotel guest book and took great delight in pointing out our entry. Westendorf was an unspoilt village when we first went there, but on our return it had become a thriving skiing centre. However, our hotel had not changed much over the years and it was good to rekindle old memories of the first fortnight of our life together.

Our first year as man and wife was spent in Edinburgh, and that period has been described in chapter 6. We found that I was the only married person on the Diploma Course and as our flat at Bruntsfield was the most spacious, it tended to be a meeting place for the Genetics students. They were probably enticed by the prospects of good food, helped down with a drink which I 'discovered' in Edinburgh, Pimms No.1. I learnt that one could progressively dilute the cocktail as the evening went on and in this way one bottle of Pimms was enough to keep a large roomful of students reasonably well satisfied. I cannot remember how I was introduced to Pimms No.1, but it has remained a favourite of mine, although more appreciated in the tropics than on a winter's evening in Edinburgh.

Bunny was also very busy during this Edinburgh year. She got a job as technical Assistant to Professors Alan Robertson and Jimmy Rendell, two eminent scientists on the staff of the Institute of Animal Genetics. Her task involved breeding populations of *Drosophila* flies which possessed one or more different genetic traits, known as mutants. Alan and Jimmy were population geneticists and their work entailed following successive generations of crossbred populations exhibiting characteristics which gave them a breeding advantage, or disadvantage. Frequently the experiments necessitated weekend work, so Bunny saw no harm in extending her lunch hour to tend our small vegetable patch. This was noticed by one of the senior staff and she was summoned to his office. The result was she was told to claim overtime for the weekend work but never to take time off in 'normal' working hours. Needless to say, we found the extra money very useful. Bunny enjoyed the work in spite of these frustrations and she liked being part of the 'Institute team', albeit in a junior capacity. Junior or not, she earned more from her employment than I did from my scholarship, so the prediction I had given to her father that she would have to support me was fulfilled – something I am not allowed to forget!

Although we both worked hard during that Edinburgh year, we obviously found time for other activities, since our first child was born in August, 1951. Knowing that my postgraduate year would end before the baby was born, we arranged that the birth would take place in London, at Queen Charlotte's Maternity Hospital, within easy reach of Bunny's old home. In those days one assumed that most births were normal and there was not the anxiety which prospective parents today seem to have about potential hazards of childbirth. It therefore came as a great shock to both of us when Bunny was rushed to Queen Charlotte's by ambulance suffering from *placenta previa*. At the time I was ignorant of this complication and was not prepared for what was turning out to be a real emergency. In those days fathers were unwelcome in maternity hospitals, so I had to wait for several hours before ringing Queen Charlotte's to learn the news. I was told I was the proud father of a baby boy and that mother and baby were doing well. When I visited Queen Charlotte's later that day, I found that the hospital's knowledge of human anatomy was a little deficient, since I was in reality the proud father of a little girl! Thankfully, mother and daughter did do well. We decided that if any return visits to maternity hospitals were required, we would try to make an appropriate booking at Queen Charlotte's. This was good long-term planning, since our next two children started their embryonic life in Uganda and Trinidad respectively. Both also needed Caesarean sections and they were all born within a few yards of each other, having started their lives several thousand miles apart.

Soon after Rosemary made her appearance, we were off on our first venture, bound for Uganda. Bunny's father told us not to worry about setting our alarm clock as he always woke up at 6 Since we did not have to leave his house in Wembley until 7.30, there would be plenty of time, he declared, to prepare for the arrival of the taxi. At 7 o'clock Rosemary woke up crying loudly, demanding to be fed! For the first time in his life, apparently, Bunny's father had overslept! However we arrived at the airport only five minutes late. In those days the planes flew from Northolt airport and the BOAC aircraft were either Hermes or Argonaults. Both these planes were luxurious compared with today's large Boeings. One even had a ladies' retiring room where babies could be fed and nappies changed. The journey from London to Kampala took three days. We landed each evening and spent the night on the ground in a convenient hotel. It was our first experience of flying and we found it very congenial. As time has gone by, flying has become increasingly uncomfortable and frustrating. Long queues must be joined at each end of the journey to check in and clear customs and immigration. On our first flight from Northolt, the taxi drove right up to the small departure lounge and after checking in, we walked straight out to the plane waiting on the tarmac a few yards away. Security checks were unknown at that time.

Our six years in Uganda have been described in chapter 7. They were happy years for me, but less so for Bunny. When we were living 'up country', I was often away on safari, and Bunny and Rosemary were left with the servants, sometimes in the middle of the bush many miles from the nearest township. Mau Mau was raging in Kenya and we never knew whether the rebellion would cross over the border. Thankfully it never did, but when the drums were beating late into the night, one never knew what information was being relayed, or whether the thinly-spread European population was about to become a target for massacre – or worse.

They were also other problems for Bunny. Whereas I had more work than I could cope with, there was very little for Bunny to do. When we were staying on the Makerere campus in Kampala, Bunny got part-time jobs demonstrating in Zoology. When we were up-country and living in Serere, she taught Agricultural Zoology to my first-year students and together we also gave a course of extra-mural classes in elementary Biology for African adults with sufficient command of English. These classes were most enjoyable because those attending were so keen and eager to learn. One wondered how far some of the mature students could have progressed had we been able to prepare them for some form of further education. But although Bunny enjoyed teaching these extra-mural students once a week, for someone who had been used to a full and interesting job, life was rather empty. She swore never again to complain about having too much to do.

Rosemary thrived throughout this period. She got on well with her Africa ayah, Anna, and she learnt to speak Luganda and Kiswahili by the age of three or four. Even when we lived for several weeks in a mud hut in the African bush, she was fit and well in spite of the fact that the hut had floors made out of cow-dung, which was 're-laid' about once a month with fresh manure. Naturally, the cow-dung crumbled as it dried out and as Rosemary was crawling at that stage, I have no doubt that several ounces of the floor covering were consumed. It never seemed to do her any harm; in fact, I believe that it helped her to develop a strong immune system which has stood her in good stead for the rest of her life. I often wonder whether the boiled, sterilised, purified and cleaned households which are so much part of life in the Western world are responsible for the rash of allergic reactions, enteric disorders and general unthriftiness of the British and American populations. In Uganda cerebral malaria and poliomyelitis were much-feared deadly diseases but, unless these potent pathogens were unwittingly unleashed on a member of the family, the more primitive life-style we all enjoyed seemed to give us lifelong protection from many of the problems encountered by subsequent generations.

By the time our first tour of duty was almost completed, there was another child on the way. Remembering our resolve to have the rest of the family come into this world at Queen Charlotte's Hospital, we immediately 'booked a bed' at the appropriate time. This time Bunny was on her own as my tour of duty was incomplete, so it was arranged that she flew home with Rosemary whilst I stayed with my students back at Serere. As before, a Caesarean section was deemed necessary. Unfortunately the combined effect of pain killers and the anaesthetic given to Bunny before the operation had a disastrous effect on baby David, who was very reluctant to start breathing. Once again, we were grateful for the skills of the doctors at Queen Charlotte's, as he suffered no permanent damage. I was given this news by Bunny's father, who, thinking he was being truthfully helpful, sent me a telegram which said 'Congratulations, its a boy; mother fine but baby poorly'. I received this news at the beginning of a bank holiday and therefore had to wait three days until I learned, to my great relief, that all was well. It was one of the worst 72 hours I have ever spent and made both of us realise how cut off we were. Bunny's father had no telephone at that time and I had to wait for a telegram, which could only be received at Serere at the end of the bank holiday. My colleagues at Serere were wonderfully supportive and for three whole evenings I was never allowed to buy a round of drinks at the Serere club house.

After our first year in Uganda, Bunny became dissatisfied with having so little to do. Even with a child to look after, most of the 'chores' were done by Anna and time began to hang a little heavily. The outcome was that Bunny decided to dismiss the cook (who, at that time, was not the best one on the station) and do the cooking herself. This in fact saved two lots of wages, as no self-respecting African cook would work in an expatriate household without the help of a 'cook's boy', whose sole function was to chop wood and keep the kitchen fire alight. I became Bunny's 'cook's boy' for a while! The quality of the meals improved dramatically, but I think we were regarded by the rest of the station as being somewhat out of the ordinary. At that time no other European wives ever went near the dirty wood-fires on which meals were cooked. However, we had a small valor stove for special dishes and for cooking cakes.

Not to be outdone, I decided to go in for a bit of landscape gardening. Most gardens had lawns flanked by flowering shrubs, such as Frangipani, Ixora and Hibiscus, but the Wilson residence sported a terrace flanked by a stone wall and rockery – quite out of the ordinary and a feature which the garden boy refused to maintain as it was 'Shauri ya Bwana', boss's work! For our final year in Uganda we returned to Kampala, where Fergus Wilson had set up a new University Farm. Our staff was now down to one house-girl, who had received no prior training. On the first day, when Bunny asked her to wash the kitchen floor, she threw a bucket of water down and swept the resultant flood out of the back door with a garden broom!

In the Autumn of 1956 I had resigned from my Lectureship at Makerere and was 'between jobs', since we were due a home leave and would not be starting work in Trinidad for five months. There were a lot of preparations to be made for this move. The most important was to pack all our personal belongings into several large wooden crates. This we did ourselves; if you wanted the job done satisfactorily, you jolly well got on and did it yourself. The boxes had to be despatched by rail to Mombassa in advance of our own journey, also by train, some days later. After our furniture had been duly crated up, Fergus Wilson, my Head of Department, and his wife Hilda kindly invited us to be their house guests for the remaining few days. The time went very quickly in a round of farewell parties on Makerere Hill. The University College was still small enough to ensure that all the expatriate staff knew one another well, in spite of the fact that some, like us, lived far away from the campus. It was a closely-knit community with a very good 'team spirit'.

The train journey to Mombassa was very eventful. The week before we left it had rained heavily on the Western scarp of the Great Rift Valley in Kenya. When we arrived at the last station in Uganda before the border, we were informed that the train could not proceed any further since a bridge had been damaged by the rains. We would have to get out and take a bus for about fifty miles and pick up another train further to the east. We were beginning to get concerned lest we should reach Mombassa after our boat, but we were told not to worry as there was plenty of time for the train to make up lost ground. We were, however, grateful that our heavy wooden crates had gone on ahead of us and should already be on board our boat.

The journey time from Kampala to Mombassa normally took two days and the train was expected to continue its journey through the night. During the afternoon the bus duly took us to the second train, now waiting in Kenya and a few hours later, a little tired and very dirty, we settled down for the night as the train chugged slowly eastwards towards the Indian Ocean. David, who at that time suffered from travel sickness, had slept all night and when I drew back the curtain from the window of our compartment to see how far we had travelled, I found that the train was where we had stopped seven hours earlier! There was, apparently, something wrong with the engine, but we were told that a replacement was on its way. I was now in something of a panic. We were at least twelve hours overdue and I was certain that, even with the fastest engine on the line, we could not reach the docks in time. I therefore persuaded the station master to send an urgent message to his colleague at Mombassa harbour station, to inform the ship's captain that about fifty passengers were still in the middle of Kenya, but doing their best to get to the ship before she sailed. I arranged for a roster of other passengers to send identical messages at every other station along the line and it was with great relief that we reached Mombassa to find the Union Castle ship *Braemar Castle* still tied up alongside the quay. I never did find out whether our numerous telegraph messages had done the trick or whether the captain could not have sailed with fifty fully paid-up passengers missing.

These various trips on the Union Castle Line to and from East Africa were memorable occasions. We were entitled to travel first class and the food was invariably excellent. How we both managed to do justice to a cooked breakfast and two other five-course main meals each day is beyond me, but in spite of five weeks of 'high living', we never put on any weight and arrived at our destination fighting fit. One of the reasons was that the ship arranged a wide variety of deck sports and on two voyages I was appointed sports' captain. This meant a lot of hard physical exercise. Another reason is that both Bunny and I seem to have sprung from genetically rather lean stock, although in recent years I appear to be making an exception to this general rule. This regular feeding was not without its amusing incidents. The Union Castle Line used to assign passengers to a set table for luncheon and dinner and once, whilst passengers on the *Durban Castle*, we had the misfortune to be saddled with a table steward who seemed to take more than his allotted rum ration before coming on duty. I started to get very irritated by the resultant poor service, but Bunny kept admonishing me, saying 'Don't make a fuss!' One evening the steward came in even more the worse for wear than usual. Shuffling up to our table, he tipped a whole dish of peas and potatoes right down the back of Bunny's dress. I looked on horrified and was about to forget the injunction not to make a fuss when Bunny rose imperiously to her feet, marched across the length of the dining room to where the head steward was presiding over the proceedings and said in a loud clear voice, 'Remove that steward. He is drunk!'. I sat open mouthed – that was my wife holding forth, who never made a fuss! The steward was promptly sent to his quarters and we never saw him again.

On the last trip home on board the *Durban Castle* we left Cape Town bound for St. Helena and hit a Force 11 gale. Ships were not fitted with stabilisers in those days and the vessel rolled from side to side alarmingly. The lounge had a grand piano, theoretically anchored to the floor by heavy bolts. However, the bolts snapped and the grand piano rushed from the port to the starboard side of the ship, smashing anything in its path. It took a gang of about a dozen strong seamen to lash the grand piano firmly to some robust fittings and thus stop the destruction which otherwise would have reduced the lounge to complete and utter chaos. While all this was going on, afternoon tea was being served in the tea lounge, the cups and saucers being placed on round tables with 'fiddles' which should have held the china and cutlery securely. However, the movement of the ship was such that the fiddles were insufficient to prevent all the china crashing to the floor in fragments. But the order was to 'serve tea', so ten minutes later the stewards swept up the broken china and laid out another service. And another; and another! After half an hour, all the cups and saucers on the ship were in smithereens and the long overdue order to 'stop serving tea' was given. We had no more fancy chinaware left for the rest of the voyage and tea thereafter was taken on crockery requisitioned from the crew's mess.

Soon after the storm had passed us by, the ship organised a fancy dress party for all the children on board. Like the other mothers, Bunny disappeared under a mountain of crepe paper and sellotape. She made a splendid dress for Rosemary, who appeared in the guise of Little Bo Peep. She played her part well and enjoyed the party. David was destined to be a Golliwog. He hadn't realised that this would entail painting his face and hands black. He was happy to put on the costume, but when the blacking appeared he was most upset. 'I don't want to be a Golliwog!' he cried. 'O yes you do,' I retorted, 'Look at all the work Mummy has put into making this splendid costume!' David lost the battle, but a very sorry looking Golliwog paraded with the other children and there was no way the judges could be expected to give him a prize for acting the part.

In spite of the storm, we enjoyed our last long trip home via the Cape of Good Hope. In many ways, the long sea voyage home was a holiday in itself. The *Braemar Castle* was an 'Intermediate' ship, which meant that it stopped, for at least a day, at every port en route. Having left Capetown, our next ports-of-call were Ascension, St. Helena, Gran Canary and up the British Channel and the Thames to London Docks. We won a competition at one of the social evenings on board ship, the prize being a 'Daimler car from the docks to any destination in the UK'. We were looking forward to this and were a little disappointed when a chap with a chauffeur's cap, but a rather grubby suit, guided us towards an ancient Austin car. On asking why we didn't have the expected Daimler, we were told, 'I am an employee of the Daimler Car Hire company, but not all the cars in our fleet are Daimlers!' However, the car was roomy and comfortable, and we were quite glad to find it had a roof-rack, which was essential since our cabin baggage was bulky and would not have fitted into the boot. Daimlers, we told ourselves, would never have been fitted with a roof rack!

Our 'home base' on all our leaves, including the period between leaving Makerere and taking up my new post in Trinidad, was Wembley. Bunny's parents were most hospitable and were more than happy to make space for our growing family for as long as we cared to stay. Bunny helped with the chores inside the house while I tried to do my bit in the garden. We always spent some time with my own parents, who had by then retired from Sanderstead to a small cottage in Wivelsfield Green in Sussex. Once when we went off on our own for a few days to visit friends, they persuaded us to leave the children with them. Unfortunately they were given bread and milk for supper each evening and this daily fare was not appreciated as much as

---

my mother would have hoped. Afterwards Rosemary said to Bunny, 'Please don't send us down there again. We just can't stand any more bread-and-milk suppers!'

One particularly happy holiday which we shared with Bunny's parents was in a farm house at Dittisham on the Dart, in Devon. The weather was superb, the farmer and his wife a very hospitable host and hostess, the food was good and everyone enjoyed themselves. Looking back, it was a unique occasion, with our little family of four and Bunny's two parents making a very happy family group. How we all managed to get into one medium-sized family car is another matter, but it was before the days of seat-belts so presumably Rosemary and David spent much of the journey on someone's lap. On several later occasions we hired a farm house or a country cottage for a week, and enjoyed a 'self catering' holiday by the sea. We developed a formula for these holidays which we followed many times with much success. Each day was allotted to one member of the family and it was their privilege to 'plan the programme', including the destination and the route to be followed as well as the activities to be pursued. This prevented a lot of potential hassle about 'Why are we going there?' or 'How much further is it to the seaside?' since when their turn came round, the children did not want their own carefully made plans to be criticised.

We also made a real effort to visit all our friends during every leave and we probably saw more of them than we would have done had we never gone abroad. It was the long periods of leave that gave us the time to 'go visiting', sometimes with and sometimes without the children. I also fitted in working visits not only to my old College at Wye, but to most of the Departments of Agriculture at all the main UK Universities. In addition I tried to get to at least one major international conference on some branch of animal science. Leaves, therefore, were very happy periods, during which our time was very fully occupied. I can almost hear Bunny muttering, as she reads this, 'So what's changed?'

Our time in Trinidad was different in many ways. Although we were still members of a minority expatriate community, Trinidad was a polyglot society, and miscegenation was commonplace. It would be wrong to say that there were no racial overtones, but these mainly took the form of political jockeying between the Trinidadians of African descent and those who had arrived from India and Pakistan after the abolition of slavery,. The ethnic mix in our time was about 40 per cent African. 40 per cent Asian, 10 per cent 'mixed' and 10 per cent the rest. The British were only a small part of 'the rest', as there were also folk from France, Spain, China and the Middle East. Not only that, but the tag of 'colonialism' did not, strictly speaking, apply solely to the British, for over its long history the Eastern Caribbean islands had been under Dutch, French and Spanish sovereignty as well. It was not forgotten by West Indian historians that the British led the move towards the emancipation of the slaves on the West Indian plantations.

For us, the most important difference was that Bunny was happy in Trinidad, where she undertook a wide variety of different jobs. When she first arrived, she was invited to assist the Zoologists by teaching Applied Entomology and helping with practical laboratory classes. Rosemary and David both attended the staff school in Trinidad. This was an 'unofficial' venture, run by the staff on a voluntary basis, the only contribution of the College being the provision of a converted building for the school premises. The school employed a full-time headmistress, who, when we arrived, was Mrs Stenson-Stenson, a somewhat formidable elderly lady who was married to the Rev. Stenson-Stenson, a retired missionary. The rest of the staff were mainly staff wives, paid more by way of an honorarium than a professional salary. Shortly after we arrived, the Chairman of the School Board, Professor John Purseglove, came round to our house and said to Bunny, 'If you don't join the school staff there will be no one to teach your daughter's class!' So Bunny became a school teacher and Rosemary and David insist that she was much stricter with them than with the other children. When it was explained to them that this was to prevent any suggestion of favouritism, which would have alienated them from their class mates, they were not entirely satisfied. 'We have to work twice as hard as anyone else to get a star!' they argued. They were probably quite right.

Mrs Stenson-Stenson had a fairly short fuse and tended to make snap decisions which she subsequently regretted and retracted. On several occasions David was promoted to Head of Class in the morning, did something wrong at lunch, and was promptly replaced, only to be reinstated at the end of afternoon school! We often laugh over this incident with the mother of his usual replacement, Martin Hodnett. In spite of these eccentricities, Mrs Stenson-Stenson was a good teacher of the old-fashioned school. She kept excellent discipline and took great care to ensure that all the children in the school were taught the three 'Rs', even if much was learnt by rote. Sadly, Mrs Stenson-Stenson died and Bunny was promoted Head Mistress of the School, but this was only for a short time as John was on the way and Bunny retired from schoolteaching in Trinidad.

When we broke the news of the new baby to Rosemary and David, Rosemary was thrilled but David was sceptical about the prospect 'I'd much rather have a puppy!' he exclaimed. We decided that he should not suffer from too much advance apprehension of the forthcoming event and a crossbred sheepdog x Labrador puppy was duly obtained. The dog was named Lassie or, to be more precise, Lassie 2nd, since we had had a previous dog of similar breeding in Uganda. Indeed, on our return to the UK, Lassie 3rd was duly procured, much to the embarrassment of Aunt Lassie, who felt, quite wrongly, that all the Wilson bitches were being named after her! The open invitation at Queen Charlotte's Maternity Hospital once more came in very useful, and John entered this world safely by way of a planned Caesarean section. Bunny was, once again, on her own at the time of the birth and I did not see my youngest son until she flew back to Trinidad with him about a month later. In those days, there was a somewhat slack procedure for clearing immigration formalities at Piarco Airport, Trinidad, and Bunny was kept waiting for quite a long time to have the baggage cleared. However, I was just the other side of a barrier separating the arrival hall from the waiting crowds of friends and relatives, so Bunny simply handed John over the barrier into the arms of his delighted big sister. I was then told, 'He's all yours now. I've done my bit!'

Soon after John was weaned, Bunny was approached by the newly arrived Director of Research at the nearby Coconut Research Station, Dr David Fenwick, who asked if she would undertake a research project for him. This work involved studying the role of a weevil as a vector responsible for the spread of a major disease of coconut palms, red ring disease. These studies kept Bunny busily occupied for about a couple of years and she was able to make some original observations, which were duly published in the form of a paper in *Tropical Agriculture* (36).

In Uganda the majority of our family holidays were taken over the border in Kenya. In Trinidad our holiday destination was nearer to home, but to reach it we had to fly from Piarco Airport, Trinidad, to Crown Point, Tobago, by DC3. For our first few holidays we hired a furnished house and did most of our own cooking. Later on we got to know the local Bishop family well. Tony Bishop was a farmer, who specialised in the production of copra from his coconut plantation and of beef cattle, which he reared on Pangola pastures planted underneath the coconut palms. When we first knew Tony, he was breeding Jamaica Red cattle, which were a cross between Red Polls imported from the UK and Zebu cattle, brought to the West Indies from India. I had seen the results of cross-breeding with Charolais stock in Latin America, where they had done very well. Tony decided to have a go, and soon he had made quite a reputation for himself for breeding some of the biggest and best beef animals on the island. We became very friendly with the Bishops. As a result, for our last few years in Trinidad, we enjoyed Tobagonian holidays staying in one of Tony's holiday cottages, idyllically situated under coconut palms, within a few yards of the sea.

Trinidad has some fine beaches but the island was formed as a result of a chunk of Venezuela being cut off from the mainland by the sea. The geology, therefore, is identical to that of mainland Latin America. Tobago, on the other hand, is mainly a coral island surrounded for the most part by a coral reef and with magnificent stretches of white sand. The reef made the swimming calm and safe and also provided a magnificent area for exploration with a snorkel. Our older children learnt to swim in these sheltered waters, and much of our holiday time was spent cruising up and down the shoreline on the landward side of the coral reef. There were sharks, but they usually stayed outside the reef. I did a bit of spear fishing and Tony Bishop took me out to sea beyond the reef, in order to fish for barracuda with the aid of a spoon attached to a fishing line. The spoon spun when dragged behind a power-boat, and looked to the barracuda like a juicy morsel, which they swallowed, only to be caught on one of several hooks attached to the line. The barracuda made excellent eating, but only if they were landed and cooked within about an hour of being caught. If the catch was large, so that several fish had to be kept in the refrigerator for another day's meal, the fish appeared bland and unappetising. I believe this is true of most fish species, in which case the majority of people have never tasted fish at their best – straight out of the water and into the pan.

Tony also introduced us to another culinary delight, roasted sweet potatoes. Sweet potatoes are usually boiled and taste quite good when cooked in this fashion, but the full flavour of a sweet potato can only be enjoyed when they are put into the embers of a fire – preferably one made from smouldering coconut shells – and slowly roasted. The resulting product is a real delight; it has a deliciously sweet, nutty taste, enhanced with thick dollops of butter, so that it literally melts in the mouth. A barbecue meal of freshly caught barracuda and sweet potatoes is one of the finest feasts I know.

When we went out for a restaurant meal, we tried as much as possible to sample local dishes. Two of these which were remarkably good were callaloo soup, made with green vegetables and crab together with anything else the cook had handy. The other dish was 'crab-back', consisting of crab meat, bread-crumbs, eggs, cheese and spices mixed together and put into a crab shell, which was in turn baked on a fire before being served. On one occasion we were having a celebratory dinner in one of our favourite restaurants in Tobago. The service was a little haphazard and when the time came for coffee and liqueurs, the waiter, who seemed to be taking part in his own private celebration, enquired what we would like to drink. Unusually, everyone ordered Drambuie and the waiter duly went off to collect them. Five minutes later he came back with an empty tray balanced somewhat precariously on his hand and said, 'I'm sorry; I didn't quite catch the order. What sort of a Buie was it you wanted?'

Throughout the West Indies the two days preceding Ash Wednesday are celebrated as 'Carnival', which, literally translated, means: 'Farewell to meat'. Carnival is taken more seriously in Trinidad than any other West Indian island. In many countries carnivals are so arranged that a small proportion of the population perform and the vast majority turn out to watch and enjoy the fun. In Trinidad, three-quarters of the population play an active part; the spectators are mainly tourists who come to see the spectacle. Carnival is preceded by several weeks of informal dancing to the music of steel bands and many house parties are arranged where the two centres of attraction are the booze (generally unlimited quantities of rum) and dancing to the music (referred to as 'Jump-ups'). By the time the actual carnival arrives, many people are already pretty party-weary, but that does not prevent the stops being pulled out for a 48-hour non-stop open-air show. The participants are grouped into various 'bands', which can be anything from a dozen people to several thousand strong, all of whom are dressed for the occasion in such a way that a common theme is enacted by all the players. Many of the bands portray historical events, such as 'Merrie England', 'Ancient splendours of China', 'The Roman Empire', etc. Other bands dress up in almost identical costumes, such as a tribe of Indians, a troupe of clowns or a group of American prisoners on death row. The members of these bands have been busy making their elaborate costumes for the previous six months, and the outfits are of very high standard and cost a great deal of money. Like wedding dresses, the costumes are never worn twice although they may be kept and passed on rather like family heirlooms. The bands are preceded by an 'orchestra', sometimes a steel band but usually a jazz band. All the groups then parade round the streets of Port-of-Spain non-stop for two days and two nights. At some point the bands wend their way round the large park in the centre of Port-of-Spain, the Queen's Park Savannah, and process across a large stage where they are 'judged' and prizes awarded for the best team effort, the best theatrical performance, the best costumes, and so on. Every year we attended these parades and they seemed to get bigger and better. The children had their own 'mini-carnival' at the Country Club and Bunny spent a lot of time making them their carnival outfits, such as Red Riding Hood for Rosemary and Noddy for David. In our last year the children (with a little help) made their own costumes and dressed up as 'Night and Day'.

Carnival was a major annual event and all the family enjoyed those days immensely.

One of the features of Carnival is that the participants are so good-natured. Although the liquor consumption is high, there are few if any incidents, but on one occasion some stupid tourists in a hired car insisted on driving down a street which the carnival procession was equally determined to process up. Not unnaturally, the insensitive car driver was subjected to a well deserved roughing-up and a few well-chosen Trinidadian insults. Although everyone was very merry, and a few were undoubtedly over the top, there was none of the usual manifestations of over-indulgence, such as street fighting, or unprovoked attacks on passers-by.

During our home leaves Bunny and I spent quite a bit of time visiting boarding schools which we thought might be suitable for the children, since it was unlikely that they would complete their secondary education in Trinidad. In order to 'express an interest', one had to pay a small 'registration fee' at each school, and we must have distributed handfuls of guineas around the public schools of England rather like confetti at a wedding. In the event, we decided that Rosemary would go to the Ashford School for Girls and this arrangement worked well. After the first few days of tearful home-sickness, Rosemary settled down and enjoyed boarding school life to the full. She made some good friends there and has kept in touch with several of them ever since. David was less fortunate. We decided, after exhaustive searching, on Caterham School in Surrey and in 1962 David duly enrolled as a boarder in the Caterham Junior School, from which he would expect to be promoted to the Senior School a couple of year's later. However, our plans did not work out as we had envisaged. Most of the junior boys were weekly boarders living locally and there were very few with parents overseas. Added to that, he was the youngest and smallest boy in the school and coming with a Trinidadian accent, he was an obvious target for any bullying. He also had to endure the coldest winter for many years, which, together with a nasty dose of 'flu, made him a desperately unhappy boy. We did not know the full extent of the problem since his letters to us were all censored and Bunny's parents thought it would be unwise to worry us since, they reasoned, there was nothing we could do about it in far away Trinidad.

By the end of David's first year, it was abundantly obvious that something was very wrong. When David first joined the school, he was usually placed in the top quartile of the form order, but after a year he was in the bottom quartile and sinking fast. Other factors came into play at that time. Bunny's father died suddenly whilst only 61, leaving a distraught widow not able to cope on her own. Lastly, the first signs of 'West-Indianisation' of the

University staff were becoming manifest and all these factors resulted in our decision to return home. Both Rosemary and David opted to leave boarding schools and go to local day-schools; Rosemary to the Dame Alice Harpur School and David to Bedford School, two of the four schools administered by the Bedford Harpur Trust. It was made clear to David that entry to Bedford School was not automatic. He had to pass a fairly stiff entrance examination. Knowing how badly he was doing at Caterham, we were somewhat apprehensive about his chances of success. 'Leave that to me,' he said. He then confidently not only passed the entrance exam but picked up a bursary into the bargain. Motivation is a very strong and powerful incentive!

John was still at the kindergarten stage on our return home and we decided to send him to Walmsley House, a well-regarded school near to our Bedford home. No entrance exam was required in his case, but he had to be approved by the Headmistress. John was very pleased with the school itself, which was far more grand than the nursery school he had previously attended in Trinidad. However, he had one major reservation. 'I won't have to work if I come here?' he asked the Headmistress. Being a wise lady, she replied, 'No, John. Not unless you decide you want to!' With that reassurance, he trotted along happily to Walmsley House for the next two years and then moved on to the Junior Department of Bedford School, known to all as 'Inky' (short for 'incubator', for reasons I never fully understood). Bunny, not to be left out on all these moves, took up a post as a Geography and Maths teacher at St. Andrew's, a girls' private school across the road from Walmsley House. She used to cycle to school with John on a 'children's carrier' attached to the bicycle and if Bunny was delayed for any reason in the afternoon, John would come across the road and be 'mothered' by the girls until Bunny was ready to leave. After a while, Bunny was made Deputy Headmistress of St. Andrew's School and was a much valued member of the staff, becoming a close friend of the Headmistress, Dorothea Pemberton, whom we still visit from time to time.

In 1968 we were on the move again, with consequent upset to the plans for the children's schooling. Rosemary and David both opted to stay on at Bedford to finish their school-days there. Rosemary had a year to go prior to taking A levels, so she lived with friends for that time. David went into the Boarding School at Bedford. John had planned to follow David to Bedford School, but preferred instead to come with us to Chester and become a day boy at King's School. As with David previously, this move necessitated passing an entrance exam, but following David's lead, he passed this with flying colours. We did not own a dog during our first years in Bedford, but John and I both missed the companionship of a canine friend and so Lassie 3rd was purchased. A further incentive to get a dog was a burglary in which Bunny's car was stolen during the night before David was due back at boarding school. Fortunately they left my company car, in which David's school bags were already loaded, but our second car was missing for a week until the police found it abandoned in Manchester.

In point of fact, we were all set to buy our next dog. However, when we turned up at the farm where she was for sale, the owner took a liking to us and we were given her free on condition we provided her with a happy home. All these three dogs were very much a part of our family, although it was always my job to train them, which I enjoyed doing as they were all so remarkably intelligent. Our houses all had well-tended flower beds and Lassie 3rd was under the impression that flower beds were a natural extension of the lawn and therefore part of the sports area when ball-games were played. I decided that this was not good for the flower beds and therefore spent time teaching Lassie that when a ball was thrown or rolled inadvertently onto a flower bed, she had to wait for someone to retrieve the ball. Within a few weeks this new skill was duly learnt but when I was not at home, if ever the ball dribbled onto a flower bed, Lassie would stop, look to see if I was around and if I wasn't, go onto the bed to fetch the ball! Dogs, they say, respect only one master and that was certainly the case with Lassie. I also taught her to walk to heel and to sit down on the pavement prior to being allowed to cross over a road. I further taught her to remain sitting on the pavement while I crossed and not to follow me until she was called. She got very good at this trick and on one occasion, having got her to sit, I crossed the road, only to run into a friend who had a long story to tell me. When he had finished, I walked on and my friend called me back. 'Isn't that your dog sitting on the other side of the road?' he asked. 'Yes', I replied, 'but she won't come until I call her!' I gave the command and Lassie gratefully joined me once more. 'I wish I could get my dog to do that,' my friend exclaimed. 'You could,' I said with my tongue in my cheek, 'but it takes a lot of careful training!'

Bunny's mother, who had stayed on at Wembley after her husband died, came up to live alongside us in a 'Granny flat' which we built as an extension onto our new house. She spent two very happy years living alongside us, but she eventually needed 24-hour nursing care and moved into the geriatric ward of a Chester hospital. When we moved from Chester to Farnham a friendly and competent geriatrician arranged for a bed to be made available in a local hospital. Bunny's mother spent her last few weeks there, where we visited her daily until she died peacefully.

Bunny had enjoyed her teaching work in Bedford and sought some similar work in Chester. She was placed on the list of 'supply teachers' for Cheshire County Council, but this meant constant changes of school and class and there was no opportunity to get to know either the pupils or the staff. Also, the nature of the work meant that one was given short notice of an assignment and this made planning of other activities tricky if not impossible. By chance, we became friendly with the solicitor through whom we had bought our Chester house. He was the Chairman of a local charity, a 'Mother and Baby' home, and he asked if Bunny would become the secretary to the management committee. Wanting something useful to do and wishing to give up supply teaching, Bunny accepted and for the next couple of years was very much involved with the administration of the home. The home had a sister in charge of the day-to-day operations and she got Bunny interested in social work, an interest which was to be taken further when we moved from Chester to Farnham.

In the spring of 1971 we had what turned out to be our last family two-week holiday together. We went to Menorca and had a wonderful time exploring the island with the help of a hired car for the first week with four nominated drivers, Bunny and I and Rosemary and David, both of whom had recently passed their driving test. In those days young drivers were not heavily surcharged for purposes of hire-car insurance and their names could be added to the policy with very little extra 'loading'. In fact most of the driving was done by Rosemary and David and this did not escape the notice of two other hotel guests in a neighbouring room, who watched with interest our two children manoeuvring the car each day with expert ease. The car was due to be returned to the hire company at the end of the first week, but our neighbours turned up one evening with a proposition. They too wanted to explore the island but could not drive and asked whether Rosemary and David would be prepared to chauffeur them if they paid for the hire of the car for a further week. Rosemary and David were delighted and they were most happy to accept the proposition with our approval. John could not join the corps of part-time chauffeurs as he was too young, so the other three of us hired bicycles for the remainder of the holidays and were able to visit parts of Menorca which cars could not reach. This unplanned arrangement for the second half of the holiday worked out very well, and the family often look back with much delight on our last holiday *en famille*.

The move from Chester to Farnham coincided with the end of school days for Rosemary and David and the commencement of their University careers. Rosemary was the first to go, leaving the Dame Alice Harpur School in Bedford to begin her studies at Liverpool University, where she took a Zoology degree. David completed his A level work a year early and opted to

take a 'year out' before going to read Physics at Leeds University. He decided
to spend a year in France in order to become fluent in at least one European
language and I managed to get him fixed up with a position in the
Computer Department of Gobain, a large French company specialising in
glass manufacture. When he told his French master at Bedford School what
he was planning to do, the master commented, 'I don't know why you want
to go to France; you were never very good at French'. 'That is precisely the
reason I have decided to go!' David replied. I was appalled at this exchange
but not all that surprised, as I had observed that foreign languages were badly
taught at school. The subject was always tackled as an academic exercise in
grammar and syntax rather than as a challenge to communicate efficiently in
another tongue. I was taught French very badly at my own school and
apparently the situation has not improved much since. This seems to be in
marked contrast to the teaching of English in most European schools, where
the students are fluent in the subject well before they reach secondary level.
Foreign languages, in my opinion, should be introduced at the primary stage,
or even better in nursery school, rather than left until students have entered
secondary school at 11 or 12 years.

I have already indicated that I first became a Rotarian in Trinidad and that,
on my return to the UK, I joined first the Bedford and then the Liverpool
Club. There was no Inner Wheel Club in Trinidad but when I transferred to
the Liverpool Club, Bunny was invited to join the Chester Inner Wheel. She
much enjoyed her time in the Chester Club and subsequently joined the
Farnham Club, where she was Secretary and then President. In 1984 she
transferred to the Edinburgh Inner Wheel Club, becoming in turn Treasurer,
President and Secretary before I joined Edinburgh Rotary, which I did only
on my retirement in 1990. Bunny's Inner Wheel activities have therefore
always been way in advance of my own Rotary achievements. At the time of
writing, I have not yet passed through the chair of any of the Rotary Clubs of
which I have been in membership. However, I have progressed to the giddy
heights of Junior Vice President of the Edinburgh Club, with the prospect of
being in the chair in two years' time.

Bunny's appetite for social work had been whetted by her stint as
Secretary of the Chester Mother and Baby Home. She did some voluntary
work when we moved to Farnham and enrolled at Guildford Technical
College in a Social Studies Course, purely out of interest. She was surprised
to find that this led to an examination for a Diploma of the University of
London which would make her degree 'relevant' for social work. Having
obtained this Diploma, she worked as an assistant social worker for Surrey
County Council based at Godalming. Her seniors persuaded her that she

60. Silver Wedding Anniversary Dinner, 1975. From left: Bunny; the author; Rosemary.

should get her 'Certificate of Qualification of a Social Worker' (CQSW). This involved becoming a postgraduate student at Southampton University for a year. I was worried that this was 'a step too far', but Bunny was determined. I believe she envied the children going off to University! On qualifying, she was welcomed back to Godalming and enjoyed her work, which she found stimulating, but also stressful at times. The area, although well known as part of the wealthy London commuter belt, had a mixed population and the usual problems that beset all strata of society.

Meanwhile, the children progressed to the next stage of their education. Rosemary, having graduated with an Honours Degree in Zoology, went on to Sheffield to take a post-graduate nursing qualification (SRN), followed by a course in midwifery. David left Bedford School and went to Leeds to read an Honours degree in Physics and then moved to Manchester Business School to take his MBA. John left King's School, Chester to enrol at the Farnham Grammar School (as it then was). He then followed his mother by going to Southampton University, graduating in Biology. So all three children followed their parents by taking up the sciences, but we never exerted any pressure on them to do so. In fact, I believe that David was hoping to do something different and take an Arts degree, but science was always the children's strongest suit and therefore won the day.

Rosemary had always said she would marry, as her mother had done, when she was 23. In 1975 she fulfilled this expectation by marrying David Lewis, a medical student. The wedding took place at St. Thomas-on-the-Bourne, in Farnham. Rosemary and David then set up home together in Liverpool and later in Stafford. Two years later Rosemary and David produced our first grandchild, Michael, when they were living in their first purchased home at Stone. Michael was followed almost two years later by James and then by Andrew. All the three boys were baptised by David's father, who had been a parish priest in Devon for many years and a canon of Exeter cathedral. David's father was very musical, playing the violin, and this musical talent clearly ran in the family, for David was a chorister at Exeter cathedral and obtained a music fellowship at Cambridge University. He has made his profession medicine, keeping his music as a well-loved hobby. All his three children inherited his vocal and instrumental abilities, specialising in clarinet, cello and organ and drums respectively. Many of our family gatherings are characterised by a musical interlude with all the Lewis family taking part, usually with the words and the scores written by David.

While Rosemary was busy bringing up her young family, our David started his first job (after gaining his MBA) with International Stores. Thereafter he has had a succession of senior positions with major national and international companies, specialising on the financial management side of their business. Finding that an accountancy qualification was sometimes deemed desirable, if not essential, in addition to the MBA, he swotted in his spare time (!) for the FCIMA qualification, which he obtained in 1995. In 1980 David announced that he was to marry his long-standing girl friend, Carol Leeming, and the wedding took place in fine style in Southampton, followed by a reception in a country hotel in the New Forest. Carol was a graduate in textile technology and has a Ph.D. degree. They first met at Leeds, where they were both students. Four years later Carol produced our first and only granddaughter, Julia, and two years after our fourth grandson, Stephen.

John had difficulty finding a job which was related to his degree in Biology, but he had always had a great interest and in-depth knowledge of computer programming and had taken computer science as a minor option in his degree course. He had spent many school holidays getting part-time jobs in computer laboratories and had a great aptitude in the subject. After many weeks applying for seemingly unavailable posts in Biology, John decided to change tack and apply for openings in the computer software world. He was successful and has stayed in that growing industry ever since. His work has taken him to Europe, firstly to Switzerland and then to the Netherlands and it was at Zeist, near Utrecht, that he met his future wife, Irene. John and Irene were married in the lovely medieval castle at Zeist and after a magnificent wedding reception, the wedding party came over to England for a 'blessing in church' at St. Thomas-on-the-Bourne. The Vicar kindly adapted the service for the benefit of the British contingent who had missed the wedding in Zeist, so that, to all intents and purposes, the service was exactly the same as if the actual wedding was being celebrated. The bride was 'given away' by her father, vows were exchanged and a 'certificate' (specially designed for the occasion) was signed and witnessed. The 'second wedding' was followed by a second reception, this time at the Bush Hotel in Farnham. A rather lovely feature of John and Irene's wedding was that David's daughter Julia was the bridesmaid and one of John's former girl friends, Carol Jeffs, was the matron of honour. Carol was by then happily married to Stephen Warnock, but both our families have been, and remain, very firm friends and we join up for combined family occasions as often as possible.

In 1990 we celebrated our Ruby Wedding Anniversary. The whole family made their way to Center Parks, in Nottingham Forest, where we enjoyed the delights on offer in the park. The younger children spent most of the time in the several large swimming pools, descending down the flumes dozens of times each day, while the others went for bicycle rides in the forest and participated in the games and sports available in the extensive grounds of the park. Rosemary and David had composed a rather moving poem about us, which was set to music and sung by the 'clan choir' after several secret rehearsals which took place when we were sent off on some pretext or other. Altogether a most memorable occasion, blessed with excellent weather and lots of good family fellowship.

The grandchildren have now progressed to the next adult generation. Michael, having obtained a degree in Medical Science from St. Andrew's University, is now completing his clinical studies at Manchester. James is at Cambridge reading history and Andrew is embarking on an engineering degree at Durham University. All three spend their vacations doing interesting things in faraway places – Malaysia, India, Latin America and the USA – and nearer home, mountain climbing, munro bagging, pot-holing and skuba diving. Julia and Stephen are still in their final years at secondary school but are equally adventurous, travelling to Europe under their own steam, staying with pen-friends and generally coming to terms with the big wide world. It is a delight to see them growing up and doing so well and one of our especial pleasures is taking them out to dinner and seeing them take an interest in good food and wine. It gives us very special pleasure when the grandchildren ring up and ask if they can come up to Edinburgh to stay for a few days, sometimes enquiring whether they could bring some friends along as well. Luckily we have three spare bedrooms which can be changed into mini-dormitories, so having a house full of lively youngsters is no problem; in fact it is a privilege and we only wish they lived nearer so that we could see more of them.

When I worked for SLF in Liverpool, the Chairman, Francis Saint, conducted my statutory Annual Appraisal. These events were common throughout Unilever at that time and are now practised almost universally in both the private and the public sector. At my first Appraisal Francis had looked at the Directors' records and said: 'I see you have only taken three weeks holiday this year.' 'Yes', I replied, 'I have had so much to do that there was no time to take the other two weeks.' Francis was unimpressed. 'Don't come back next year with the same excuse. Unilever does not give us five weeks holiday because it is generous; it gives us five weeks because we all need it in order to do our jobs properly for the rest of the year!' This was a very salutary piece of advice, which I have heeded ever since. When I retired in 1990, I decided to increase my annual 'holiday allowance' from the five weeks I was instructed to take in 1968 to eight weeks. Each year Bunny and I plan a holiday schedule which nothing is allowed to disrupt and we have greatly enjoyed the 'time and space' for being alone together and enjoying places we had not visited heretofore. These holidays have included cruising up the Nile in 1990, cruising up the Norwegian Fjords in 1991, sailing round the British Isles on a Bird Watching cruise with the RSPB in 1992, visiting Australia and New Zealand in 1993, motoring up the Canyons of the USA in 1994, visiting Peru and Ecuador and the Galapagos Islands in 1995, visiting Canada with the help of Toronto-based friends in 1996, cruising with the Scottish National Trust in 1997 and 1998, cruising up the Yangtze river in China in 1998, touring northern India in 1999 and cruising the Eastern Mediterranean and the western isles of Scotland in 2000. We have made many new friends on these holidays. We keep in regular touch with many of them, especially Eddie and Fran Gaylor (whom we met on a holiday to Jugoslavia), Derek and Peggy Bantock (with whom we have twice shared Italian holidays) and Mike and Mary Bennett (whom we met in Canada and have since visited on the trip to the American Canyons in 1994). In addition, we spend at least another four weeks each year visiting the family in England and Holland and exploring parts of the British Isles hitherto unknown to us. There is only one disadvantage in these most enjoyable times away from home and that is the mountain of work awaiting us on our return! But the holidays are fun. More than that, they are an essential part of our retirement and we enjoy them to the full.

61. Golden Wedding Anniversary Lunch, Frensham Ponds, Surrey, 2000.
The author with his wife, Bunny.

In September, 2000, we celebrated our Golden Wedding. In fact we had a double celebration, the first stage being in Surrey, not far from where we were married and even nearer to the church at Farnham where Rosemary had her wedding twenty-five years ago. We therefore combined our Golden with Rosemary's silver wedding and we were delighted that we were able to share the occasion with all the members of our immediate family. Two weeks later, Bunny and I held a smaller dinner party, for 50 guests, in Edinburgh when we were able to entertain many of our Scottish friends who had not been able to make the journey to Surrey. Not to be outdone, John and Irene have marked twelve and a half years of happy marriage. This particular anniversary, unknown in the UK, is taken very seriously in the Nederlands and all our family went to Zeist in February to celebrate this important occasion in their lives. It is a great pleasure and privilege to be members of our own very special 'extended family'. It is the many joys which we share with our children and grandchildren that are the hallmark of our own particular pattern of 'Married Bliss'.

References

1. Wilson, P.N. (1952) Growth analysis of the domestic fowl. I. Effect of plane of nutrition on growth rate and external measurements. J.Agric.Sci. 42, 369-381.
2. Wilson, P.N. (1954) Growth analysis of the domestic fowl. II. Effect of plane of nutrition on carcass composition. J.Agric.Sci. 44, 67-85.
3. Wilson, P.N. (1954) Growth analysis of the domestic fowl. III. Effect of plane of nutrition on carcass composition of cockerels and egg yields of pullets. J.Agric.Sci. 45, 110-124.
4. Wilson, P.N. & Watson, J.M. (1955) Two surveys of Kasilang Erony, Teso, 1937 and 1953. Uganda J. 20, 182-197.
5. Wilson, P.N. and Watson, J.M. (1955) The flora of Kasilang Erony, Teso. Uganda J. 19, 183-193.
6. Wilson, P.N. (1957) Further notes on the development of the Serere herd of Shorthorned Zebu cattle. Emp.J.Exp.Agric. 25, 263-277.
7. Wilson, P.N. (1957) Effect of contrasted planes of nutrition on the carcass development and composition of the East African Dwarf Goat. Nature (Lond.) 180, 145-146.
8. Wilson, P.N. (1957) Studies on the browsing and reproductive behaviour of the East African Dwarf Goat. E.Afric.Agric.J. 23, 138-147.
9. Wilson, P.N. (1958) The effect of plane of nutrition on the growth and development of the East African Dwarf Goat. I. Effect on the liveweight gains and the external measurements of kids. J.Agric. Sci. 50, 198-210.
10. Wilson, P.N. (1958) An agricultural survey of Moruita Erony, Teso, Uganda. Uganda J. 22, 22-38.
11. Wilson, P.N. (1958) The effect of plane of nutrition on the growth and development of the East African Dwarf Goat. II. Age changes in the carcass composition of female kids. J.Agric.Sci. 51, 4-21.
12. Wilson, P.N. (1958) Animal Husbandry – the problem child of Trinidad Agriculture. J.Agric.Soc.Trin. & Tob. 58, 425-467.
13. Wilson, P.N. (1959) Cattle Breeding in Trinidad. J.Agric.Soc.Trin. & Tob. 59, 451-454.
14. Wilson, P.N. (1959) Animal Husbandry investigations at ICTA with special reference to the management of cattle on Pangola pastures. J.Agric.Soc.Trin. & Tob. 59, 455-468.
15. Wilson, P.N. (1960) The effect of plane of nutrition on the growth and development of the East African Dwarf Goat. III. The effect of plane of nutrition and sex on the carcass composition of the kid at two stages of growth. J.Agric.Sci. 54, 105-130.
16. Wilson, P.N. (1960) Pasture Grass Investigations in Trinidad with special reference to Pangola Grass (*Digitaria decumbens*) Proc. 8th. Int.Grassl.Cong., 390-392.
17. Osbourn, D.F. and Wilson, P.N. (1960) Effects of different patterns of allocation of a restricted quantity of food upon the growth and development of cockerels. J.Agric.Sci. 54, 278-289.
18. Wilson, P.N. and Osbourn, D.F. (1960) Compensatory growth after under-nutrition in mammals and birds. Biol.Rev.35, 324-363.
19. Adeniyi, S.A. and Wilson, P.N. (1960) Studies on Pangola grass at ICTA. Effects of Fertiliser Application at time of establishment, and cutting interval, on the yield of ungrazed Pangola Grass. Trop.Agric. (Trin.) 37, 271-282.
20. Wilson, P.N. (1960) Grass – the World's most important crop. J.Agric.Soc. Trin.& Tob. 60, 483-521.
21. Wilson, P.N. and Herrera, E. (1961) List of completed research work carried out at the ICTA on animal and grassland husbandry. J.Agric.Soc.Trin. & Tob. 61, 63-86.

22. Wilson, P.N. (1961) Observations on the grazing behaviour of cross-bred Zebu Holstein cattle managed on Pangola pastures in Trinidad. Turrialba, 11, 57-71.

23. Butterworth, M.H., Groom, C.G. and Wilson, P.N. (1961) The intake of Pangola grass (*Digitaria decumbens*) under wet and dry season conditions in Trinidad. J.Agric.Sci., 56, 407-410.

24. Wilson, P.N. (1961) Agricultural Research and the University. Univ.Quarterly 15, 276-281.

25. Wilson, P.N. (1961) The grazing behaviour and free water intake of East African short-horned Zebu heifers at Serere, Uganda. J.Agric. Sci. 56, 351-364.

26. Wilson, P.N. (1961) Palatability of water buffalo meat. J.Agric.Soc.Trin. & Tob. 61, 457-460.

27. Wilson, P.N. (1961) Agricultural education in the West Indies. J.Agric.Soc. Trin. & Tob. 61. 461-488.

28. Wilson, P.N. (1961) Animal Breeding in the Tropics. Rep. of Conf. of Directors of Agric., Wye College, London, Sept. 1961.

29. Wilson, P.N., Fewkes, D.W. & Emsley, M.G. (1962) Note on a heavy infestation of Pangola Grass (*Digitaria decumbens*) by the sugar-cane froghopper (*Aeneolamia varia saccharina*). Trop.Agric. 39, 49-51.

30. Wilson, P.N., Barratt, M.H. and Butterworth, M.H. (1962) The water intake of milking cows grazing Pangola grass under wet and dry season conditions in Trinidad. J.Agric.Sci. 58, 257-264.

31. Wilson, P.N. and Houghton, T.R. (1962) The development of the herd of Holstein Zebu cattle at the ICTA, Trinidad. Emp.J. Exp.Agric. 30, 160-180.

32. Wilson, P.N. (1962) The Imperial College of Tropical Agriculture – Past, Present and Future. Caribbean Agric. 1, 1-10.

33. Wilson, P.N. (1963) The need for more research on the economics of grassland and livestock production in the Caribbean. Caribbean Agric. 1, 95-110.

34. Wilson, P.N. (1963) The Teaching and Research programme of the Department of Agriculture of U.W.I. J.Agric.Soc.Trin. & Tob. 63, 285-302.

35. Wilson, P.N. (1963) Agricultural Prospects for Trinidad and Tobago. J.Agric. Soc. Trin. & Tob. 63, 473-494.

36. Wilson, M.E. (1963) Investigations into the development of the Palm Weevil *Rhynchophorus palmarum (L.)* Trop.Agric. (Trin.) 40, 185-196

37. Wilson, P.N. & Osbourn, D.F. (1964) Experimental work on Pangola Grass (*Digitaria decumbens*) at the ICTA, Trinidad. Surinamse Landbouw Bull. 82, 173-191.

38. Wilson, P.N. (1965) Carcass assessment and meat quality. Occasional Symp.Proc. 2, Brit. Grassl.Soc. 83-90.

39. Webster, C.C. & Wilson, P.N. (1966) Agriculture in the Tropics 1st edn. Longmans Green, Lond.

40. Wilson, P.N. & Dew, R.E. (1967) Variation and definition of the killing-out percentage of beef cattle. Proc. CICRA Cong. Dublin, May 1967. p. 78.

41. Wilson, P.N. (1967) Carcass quality with lambs. The effect of normal rearing, early weaning and artificial rearing on fat quality. Proc.CICRA Cong.Dublin, May 1967. 88-89.

42. Wilson, P.N. (1968) Biological Ceilings and economic efficiencies for animal production. Chem. & Ind. 27, 899-902.

43. Wilson, P.N. (1971) New cereal varieties and the Feed Manufacturer. Milling, 153, 21-26.

44. Wilson, P.N. (1971) Meat Production in Developed and Developing Countries: Present and Future. Proc. IFST 4(3), 91-96.

45. Wilson, P.N. (1972) The need for intensification in animal production and the subsequent pollution problem. J.Sci.Fd. & Agric. 23, 1393-1398.

46. Wilson, P.N. (1972) Back to nature for good housing. Livestock Fmg. August, 1972.

47. Wilson, P.N. (1973) The Biological Efficiency of Protein Production by Animal Production Enterprises. In 'The Biological Efficiency of Protein Production'. (ed. J.G.W. Jones) C.U.P. pp. 201-209.

48. Cuthbert, N.H., Thickett, W.S. and Wilson, P.N. (1973) The effect of varying protein level in a ꞏ

49. Wilson, P.N. (1973) Livestock Physiology and Nutrition. Phil.Trans. Roy.Soc.Lond. B 267, 101-112.

50. Wilson, P.N. (1974) Future Feeds for Livestock. Span, 17, 8-10.

51. Wilson, P.N. (1974) Development of Animal Sciences at ICTA. Trop. Agric. (Trin.) 51, 491-495.

52. Wilson, P.N. (1974) Possibilities for processing straw for animal feed. Rep. of the MAFF/DAFS straw Utilisation Conf. Oxford (ed. A.R.Staniforth) 19-24.

53. Forster, N.C., Wilson, P.N. and Dixon, T.W. (1974) Advances in computerised dairy management aids. Proc. Br. Soc. Anim. Prod. 3, 105. (Abstr.)

54. Wilson, P.N. (1974) Animal feeding AD 2000. In Symp.for 125th. Anniv. of Koninklijke J.Willebeek Le Mair & Co. BV Rotterdam, May, 1974 pp. 10-30.

55. Wilson, P.N. (1975) Feed v Food. Unilever International, Sept. 1975, pp. 34-39.

56. Wilson, P.N. (1975) The Compound Industry in Animal Production. Span, 18, 117-118.

57. Wilson, P.N. (1975) General Considerations. Proc. BSAP. Symp. on Cattle Experimentation. ed.Wood, P.D.P. 19-33 Dec. 1975.

58. Wilson, P.N. & Jones, G. (1975) Contributions of medicinal Feed Additives to Agriculture. In 'Medicinal Feed Additives for Livestock Proc. AVI Symp., Unwin, Lond. 1-18

59. Wilson, P.N. (1976) Optimum size and structure of enterprises. In 'Meat Animals' (eds. Lister D. et. al.), Plenum Publ. New York. 57-68.

60. Wilson, P.N. (1976) Present and Future sources of Protein. In 'International Grain and Feed Markets Forecast and Statistical Digest'. Turret Press, Lond. 34-36.

61. Wilson, P.N., Sangster, I. & Walker, C. (1976) Experience in running a commercial straw processing plant. Rep.MAFF/ADAS Straw Utilisation Conf. (ed. A.R.Staniforth) Oxford, 8-12.

62. Wilson, P.N. (1976) Nutritionally improved straw. Agric.Merchant. 56, 41-46.

63. Wilson, P.N. (1976) Agricultural development from a commercial viewpoint. In 'Agricultural Development. SADC Seminar Rep. on the need, definition, functions and implications in Scotland'. pp. 21-40.

64. Wilson, P.N. (1976) Practical aspects of implementing a comprehensive metabolic profile advisory service for dairy cows. Proc.3rd. Int.Conf. on Prod.Diseases in Farm Animals. Wageningen, Holland. pp. 50-56.

65. Wilson, P.N. (1976) Input/Output relationships of dairy cows with particular reference to the law of diminishing returns. Proc. 3rd. Int.Conf. on Prod.Diseases in Farm Animals. Wageningen, Holland. pp. 56-61.

66. Wilson, P.N. (1976) Grasping at Straw in Animal Feedingstuffs. Agric.Merchant, Nov. 1976 pp. 41, 45-46.

67. Wilson, P.N. (1977) Man and Animal: The Conflict. Hereford Breed J. IX, 3, 137-139.

68. Cuthbert, N.H., Thickett, W.S., Wilson P.N. & Brigstocke, T.D.A. (1967) The use of Sodium Hydroxide treated straw in rations for beef cattle. Anim. Prod. 24, 1547 (Abstr.)

69. Wilson, P.N. (1977) Straw – the new energy cattle feed. In 'Science into Practice', Farmers Club J. 26, 34-62.

70. Wilson, P.N. (1977) Future prospects for the Jersey. The Jersey, 121, 37-39.

71. Wilson, P.N. & Brigstocke, T.D.A. (1977) The use of NaOH-treated straw (NIS) in rations for dairy cows – an interim report of recent work. Rep. of MAFF/ADAS Straw Utilisation Conf. (ed. A.R. Staniforth), Oxford, pp. 39-41.

72. Wilson, P.N. & Brigstocke, T.D.A. (1977) Recycling farm waste for Animal Feed. Agric.Prog. 52, 49-58.

73. Wilson, P.N. (1977) The Composition of Animal Feeds. J.Sci.Fd.Agric. 28, 717-727.

74. Wilson, P.N. & Brigstocke, T.D.A. (1977) The Commercial Straw Process. Process Biochem. 12, 17-21.

75. Wilson, P.N. (1977) Where stands meat in the competition between human food and animal feed? Vet.Rev. Supplement to Vet. Drug., Sept. 1977. pp. 3-4.

76. Wilson, P.N. & Brigstocke, T.D.A. (1977) Energy and UK Agriculture. Long Range Planning 10, 64-70.

77. Wilson, P.N. (1977) World Trends in Beef Production. In CENTO Seminar on Recent Technology on Beef production: Breeding, Feeding and Management in Intensive and Semi-Intensive Systems. CENTO, Ankara, Turkey. pp. 4-33.

78. Wilson, P.N. (1977) Inter-relationship between nutrients, feed, milk and money. In Preventive Medicine in Bovine Practice. Proc.Brit.Cattle Vet.Assoc.Conf. pp. 24-36

79. Wilson, P.N. (1977) British Agriculture – a contradiction of threats and challenges. Agric.Merchant, 57, 10-11 and 16 and 40.

80. Wilson, P.N. & Brigstocke, T.D.A. (1978) A new Fertility Technique. Unilever Magazine, 25, 19-21.

81. Wilson, P.N. (1978) Metabolisable Energy (ME) – Theory and Practice. Agritrade May 1978. pp. 57-67.

82. Wilson, P.N. & Brigstocke, T.D.A. (1978) The role of Crude Fibre in Diets for Diary Cows. Livestock Farming, May, 1978. pp. 59-60.

83. Wilson, P.N. (1978) Nutritionally Improved Straw – a Qualitative and Quantitative Assessment. In Straw – An Asset or a Liability to the Farmer? Abstr. NFU Conf. April, 1978.

84. Wilson, P.N.& Brigstocke, T.D.A. (1978) Animal Nutrition over 20 years. Span, 21, 66-68.

85. Wilson, P.N. (1978) Perspectives for Milk Production – UK v USA. Rep. of Edinburgh Dairy Conf. on Breeding, Feeding and Management of High Yielding Herds. Edin.School of Agric. Sept, 1978 pp. 1-22

86. Wilson, P.N. (1978) The role of Commercial Organisations. Agric. Progress 53, 22-28.

87. Wilson, P.N. & Brigstocke, T.D.A. (1978) Complete Diet Feeds – a mixture of views. What's new in Farming 2(1), October, 1978. pp. 18 -23

88. Wilson, P.N. & Medd, R.K. (1978) Blood Profiles as a guide to nutritional status in the field. In The use of Blood metabolites in Animal Prod., BSAP Occ.Publ. no.1. pp. 155-162

89. Wilson, P.N., Brigstocke, T.D.A. & Cooke, B.C. (1978) Copper in the diet of Pigs. AgriTrade. August, 1978. pp. 38-39, 41, 54

90. Cuthbert, N.H., Thickett, W.S., Wilson, P.N. & Brigstocke, T.D.A. (1978) The use of sodium hydroxide treated straw in rations for beef cattle. Anim.Prod. 27, 161-169.

91. Wilson, P.N. & Brigstocke, T.D.A. (1978) The Sheep Market – Meat and Wool as it affects UK farmers. Farm Management 3, 465-476.

92. Wilson, P.N. (1978) The Cattle Industry: Recent Developments and Future Welfare Requirements. In The Welfare of Food Animals. Proc. UFAW Symp. pp. 9-23.

93. Wilson, P.N. (1979) A Guide to providing the Farmer with more meaningful Nutritional Information: An Analysis. AgriTrade. March , 1979. pp. 24-27, 30, 34-35

94. Wilson, P.N. (1979) Concentrates. In Feeding Strategy for the High Yielding Diary Cow. (eds W.H.Broster & H.Swan). Crosby Lockwood Staples, London. EAAP Publ. no. 25. pp. 374-397

95. Brigstocke, T.D.A., Thickett, W.S., Cuthbert, N.H., Wilson, P.N. & Lindeman, M.A. (1979) Comparison of a standard dairy cow diet with nil and 40 % cassava, fed in conjunction with a high intake of grass silage. Anim.Prod. 28, 413 (Abstr.)

96. Thickett, W.S., Cuthbert, N.H., Brigstocke, T.D.A. and Wilson, P.N. (1979) The inter-relationship between liveweight gain, performance, water intake and colostrum status in calf early-weaning systems. Anim.Prod. 28, 419 (Abstr.)

97. Cowdy, P.E.M., Eddie, S.M., Brigstocke, T.D.A. and Wilson, P.N. (1979) The inclusion of ground lupin seed on pig performance in a standard fattening ration. Anim.Prod. 28, 433 (Abstr.)

98. Wilson, P.N., Brigstocke, T.D.A. and Cooke, B.C. (1979) Copper as an inexpensive growth promoter for the pig. Process Biochem. 48, 10-16.

99. Thickett, W.S., Wilson, P.N., Cuthbert, N.H., Brigstocke, T.D.A. & Lindeman, M.A. (1979) An evaluation of 45% NaOH-treated straw diet fed ad lib in conjunction with grass silage to beef cattle. In Forage conservation in the 1980's. Occ.Symp.No. 11 (ed. C.Thomas) Eur. Grassld. Fed. Publ. pp. 424-427.

100. Cooke, B.C., Filmer, D.G., Wilson, P.N., Hall, G.R., Speight, D. and Roberts, P. (1979) Some aspects of a survey on the growth response of pigs to the dietary inclusion of copper. Anim. Prod. 28, 136, 137

101. Wilson, P.N. (1979) Rearing the Dairy Heifer. In World Jersey Cattle Bur. Conf. Proc. pp. 4, 1-5.

102. Wilson, P.N. (1979) The principles of Complete Feeding for Dairy Cows. In Cattle Feeding, Handling, Fertility and Preventive Medicine. Proc.Conf. Brit.Cattle Vet.Assoc. Seale Hayne College, April, 1978. pp. 74-89

103. Wilson, P.N. & Brigstocke, T.D.A. (1980) Energy Usage in British Agriculture – A Review of Future Prospects. Agric.Systems 5, 51-70.

104. Wilson, P.N. (1980) The use of By-Products in Animal Feeds. In By-Products and wastes in Animal Feeding. (ed. E.R.Orskov) Occ.Publ. No.3. Brit.Soc.Anim.Prod. 1980. pp. 113-117.

105. Thickett, W.S., Cuthbert, N.H., Brigstocke, T.D.A., Wilson, P.N. & Lindeman, M.A. (1980) Performance and management of calves reared on cold acidified milk replacers fed *ad lib.* Anim.Prod. 30, 459 (Abstr.)

106. Cuthbert, N.H., Crookes, F.G., Wilson, P.N., Brigstocke, T.D.A. & Dixon, T.D. (1980) Demonstration of alternative feeding systems as a basis for extension. Anim.Prod. 30, 472-473.

107. Wilson, P.N. (1980) R&D contribution to broiler nutrition and promising areas for future research. Paper presented at AL, Oslo Symp.1980. 14 pp.

108. Wilson, P.N., Brigstocke, T.D.A. and Williams, D.R. (1980) The feeding of poultry waste to ruminants. Process Biochem. 15, 36-41 & 48.

109. Brigstocke, T.D.A., Wilson, P.N. & Cuthbert, N.H. (1980) Feeding and management of the ewe at lambing. The Sheep Farmer, Spring 1980 pp. 33, 35-36, 38 & 39.

110. Webster, C.C. & Wilson, P.N. (1980) Agriculture in the Tropics (2nd. edn.) Longmans, London.

111. Wilson, P.N. & Strachan, P.J. (1981) The contribution of ungraded protein to the protein requirements of dairy cows. In Recent Advances in Animal Nutrition (ed. W. Haresign). Butterworths, London pp 99-118

112. Wilson, P.N., Brigstocke, T.D.A. & Cuthbert, N.H. (1981) Some factors affecting the future composition of Compound Animal Feed. Anim.Feed Sci. & Technol. 6, 1-14.

113. Wilson, P.N. & Brigstocke, T.D.A. (1981) The relevance of the animal in vegetable production systems. In Vegetable Productivity (ed. C.R.W.Spedding) Symp. Inst. Biol. no 25, Macmillan Publ.Ltd. pp. 50-73.

114. Brigstocke, T.D.A., Lindeman, M.A., Cuthbert, N.H., Wilson P.N. & Cole, J.P.L. (1981) The dry matter intake of Jersey cows. Anim.Prod. 32, 366 (Abstr.)

115. Lindeman, M.A., Brigstocke, T.D.A. & Wilson, P.N. (1981) The response of growing rabbits to varying levels of sodium hydroxide treated straw. Anim.Prod. 32, 389-390 (Abstr.)

116. Lindeman, M.A., Brigstocke, T.D.A., Cuthbert, N.H. & Wilson, P.N. (1981) The application of the ARC Protein System on Farm. In Energy and Protein Requirements of Ruminants. Proc. 4th. Study Conf. of the Scottish Agricultural Colleges, pp. 60-73.

117. Wilson, P.N., Brigstocke, T.D.A. & Robb, J. (1981) Nutritionally Improved Straw (sodium hydroxide treated straw)., AgriTrade July, 1981, pp. 26-32.

118. Thickett, W.S., Cuthbert, N.H., Brigstocke, T.D.A., Lindeman, M.A. & Wilson, P.N. (1981) The management of calves on an early weaning system. 1. The relationship of voluntary water intake to dry feed intake and liveweight gain to 3 weeks. Anim.Prod. 33, 25-30.

119. Brigstocke, T.D.A., Cuthbert, N.H., Thickett, W.S., Wilson, P.N. & Lindeman, M.A. (1981) A comparison of a dairy cow compound feed, with and without cassava, given grass silage. Anim.Prod. 33, 19-24.

120. Wilson, P.N. & Brigstocke, T.D.A. (1981) The contribution of medicinal feed additives to livestock production. Feed Compounder, Dec. 1981. pp. 7-10, 12.

121. Wilson, P.N. (1981) The viewpoint of the feed compounder. Summary of paper presented at an Agric.Group.Symp. on the ARC Recommendations for the Protein and Major Mineral Requirements of Ruminants, 1980. J.Sci.Fd.Agric. 32, 986-987.

122. Wilson, P.N. & Brigstocke, T.D.A. (1981) Energy utilisation in orthodox and biological agriculture: A comparison. In Biological Husbandry: A scientific approach to organic farming (ed. B.Stonehouse). Butterworths, London. pp. 301-317

123. Wilson, P.N. (1981) Animal Welfare. Span. 24, 121-123.

124. Brigstocke, T.D.A., Lindeman, M.A., Cuthbert, N.H., Wilson, P.N. & Cole, J.P.L. (1982) A note on the dry matter intake of Jersey cows. Anim.Prod. 35, 285-287.

125. Lindeman, M.A., Brigstocke, T.D.A. & Wilson, P.N. (1982) A note on the response of growing rabbits to varying levels of sodium hydroxide treated straw. Anim. Prod. 34, 107-110.

126. Wilson, P.N. & Brigstocke, T.D.A. (1982) The improved feeding of Cattle and Sheep. Granada Technical Books, London.

127. Wilson, P.N. (1982) Open Declaration – the Compounders' case. AgriTrade March, 1982. pp. 24-25.

128. Wilson, P.N. & Brigstocke, T.D.A. (1982) Factors affecting meat production in the future. The Agric. Engineer, Summer, 1982. pp. 35-40.

129. Wilson, P.N. (1982) Animal Welfare – the real issues that concern us. AgriTrade August 1982. pp. 28-29.

130. Wilson, P.N. (1982) NIS – a new animal feed from alkali treated straw. UNEP Industry & Environment, Spring 1982. pp. 23-26.

131. Thickett, W.S., Cuthbert, N.H., Brigstocke, T.D.A., Lindeman, M.A. & Wilson, P.N. (1983) A note on the performance and management of calves reared on cold acidified milk replacer fed *ad lib*. Anim.Prod. 36, 147-150.

132. Wilson, P.N., Brigstocke, T.D.A. & Williams, D.R. (1983) The new EEC Compound Feed Directive (79/373) with special reference to changes from existing UK legislation. Conf. Soc.of Feed Technol., London, 1983.

133. Wilson, P.N. & Brigstocke, T.D.A. (1983) Economic considerations. In Upgrading Wastes for Feed and Food. Proc 36th Easter School of Agric.Sci, Univ. of Nottingham. Butterworths, London, pp. 291-306.

134. Wilson, P.N. (1983) The Resources of Agriculture. Feeds for Animals (including imported and manufactured products). In Fream's Elements of Agriculture (ed. C.R.W.Spedding) Murray, 1983.

135. Wood, P.D.P. & Wilson, P.N. (1983) Some attributes of very high yielding British Friesian and Holstein Dairy Cows. Anim.Prod. 37, 157-164.

136. Wilson, P.N. & Wood, P.D.P. (1983) Some nutritional aspects of high yielding cows. In Proc.16th. Feed Manufacturers Conf. Nottingham Univ., Butterworths, London, pp. 179-197.

137. Wilson, P.N. & Lawrence, A.B. (1984) Research and development implications for the future. In Milk Compositional Quality. Occ.Publ.No.9, Brit.Soc. Anim.Prod. pp. 95-106.

138. Wilson, P.N. (1984) The economics of the C.I. Breeds – Future prospects for management, feeding and genetic improvement. Gold Top News, Nov.1984, pp. 7-10.

139. Lowman, B.G., Wilson, P.N. & Lawrence, A.B. (1984) Housing, Health and Management in the Beef Breeding Herd. Res. & Dev. in Agric. 1, 181-186.

140. Wilson, P.N. (1985) A Review of the Industry. The Feed Compounder, 5, 28-30.

141. Wilson, P.N. & Lawrence, A.B. (1985) Fats in compound feeds – a Review. Chem.Ind. 4, 113-118.

142. Wilson, P.N. & Lawrence, A.B. (1985) Animal Husbandry; the period 1973-1995. Phil. Trans. Roy.Soc.Lond. B 310, 275-288.

143. Wilson, P.N. & Simm, G. (1985) What do breeders want? Proc.Brit.Cattle Breeders Club Digest 40, 41-46.

144. Wilson, P.N., Emmans, G.C. & Lawrence, A.B. (1985) Improved energetic and biological efficiencies of meat vs crop production. Centre for Human Ecology Workshop, Edinburgh, May 1985.

145. Wilson, P.N., Lawrence, A.B. and Appleby, M.C. (1985). European View of Animal Welfare. Proc.Int.Annual Meeting Anim.Health Inst., Florida, USA, Dec. 1985.

146. Wilson, P.N. (1985) Animal Production – the next decade. J.Edin.Agric.Former Students Assoc. 59, 2-12.

147. Wilson, P.N. (1986) Agricultural Animal Welfare: The European View. Feed Management 37, 14-20.

148. Wilson, P.N. (1986) Gulls in Research and Development Scottish Farming Leader, Feb.1986, pp. 12-13.

149. Wilson, P.N. (1986) The challenge facing the poultry industry. AgriTrade, Jan. 1986 pp. 10-12.

150. Wilson, P.N. (1986) The future prospects for the Scottish Agricultural Colleges in research, teaching and extension. Proc.Scottish Agric. Arbiters' Assoc. 1986, pp. 20-36.

151. Wilson, P.N. & Lawrence, A.B. (1986) The future of minority breeds. The Ark 8, 237-240.

152. Wilson, P.N. (1987) Farms, Woods and Forests. Scottish Farming Leader, July 1987. pp. 15-17.

153. Wilson, P.N. (1987) The Review of the Poultry Industry for 1986. Feed Compounder, January, 1987. pp. 10-11.

154. Wilson, P.N. (1987) Animal Nutrition; the contribution of Chemistry. Chem. in Brit. 23, 335-340.

155. Wilson, P.N. (1987) Biology in Animal Production. Proc.Inst.Biol. Symp.Antrim, N.I., pp. 1-16.

156. Wilson, P.N. & Lawrence, A.B. (1987) Changes in livestock nutrition and husbandry. In Farming into the Twenty First Century. (Ed.J.Glasser) Conf.Rep.Norsk Hydro Fertilisers, pp. 118-145.

157. Wilson, P.N. (1987) The challenge to the animal feed compounding industry. Proc.Nat.Poultry Conf. Lincoln, March 1987. 12 pp.

158. Wilson, P.N. (1987) New Product Innovation. Proc.Assoc.Vet.Ind.Conf. London, Nov. 1987 9 pp.

159. Wilson, P.N., McKinlay, R.G. & Loman, B. (1989) The role of agro-chemicals in animal and crop production. Chem. & Ind. 2, 29-35.

160. Wilson, P.N. (1988) Major factors impacting on the compound feed producer over the next five years. Proc.12th. Feed Ind.Conf, Nottingham, October 1988.

161. Wilson, P.N. (1988) The future of world agriculture: The food supply gap that's getting wider. Modern Farming, 5, 2-10.

162. Wilson, P.N. (1988) The future role of feed additives. Feed Compounders' Conf. Maidenhead, June, 1988, 11 pp.

163. Wilson, P.N. (1988) The trend towards leaner meat and low fat milks: implications for agriculture. Proc.MRC Conf, Cambridge, July, 1988. 14 pp.

164. Wilson, P.N. & Dun, P. (1988) Overview of the poultry industry. Proc.RASE Overseas Poultry Conf., Stoneleigh, May 1988. 9pp.

165. Wilson, P.N. (1989) BST – The Facts. Feed Compounder 9, 14-17.

166. Wilson, P.N. (1989) Food Requirements and Agricultural Implications for the next Millennium. Proc.Inst.Biol.Conf. on Food & Farming, London, April 1989.

167. Wilson, P.N. (1989) The greenhouse effect: effects on animal production in the UK. Conf.Roy.Soc. London, July, 1989. 19 pp.

168. Wilson, P.N. (1989) The Edinburgh School of Agriculture: Past, Present and Future. Scottish Farmer, 3rd and 10th June, 1989 pp. 11-12 & V1-V111.

169. Wilson, P.N., Bourchier, C.P., Kelly, T.J., Leaver, J.D. & Spring, D.G. (1990) The Wilson Report. mimeo MMB, Thames Ditton. 76 pp.

170. Fleming, I.J. & Robertson, N.F. (1990) Britain's First Chair of Agriculture at the University of Edinburgh, 1790-1990. East of Scotland Coll.of Agric., Edinburgh. 114 pp.

171. Wilson, P.N. (1990) Edinburgh celebrates 200 years of agricultural education. Farmers Club J., 108, 22-27.

172. Wilson, P.N. (1990) Agriculture – the next 15 years. 2nd. Int.Sci.Festival Guide, 10-11.

173. Coutts, B. (1991) Bothy to Big Ben. Mercat Press. 115 pp.

174. Percy, F.H.G. (1991) Whitgift School – A History. The Whitgift Foundation. 352 pp.

175. Boschoff, W. (1992) My African Notebook. Print Studio (Pty) Ltd. 102 pp.

176. Richards, S. (1994) Wye College and its World. Wye College Press. 336 pp.

177. Wilson, P.N. (1995) Growth and Development of the scientific qualities of the British Society of Animal Production. Anim.Sci. 60, 1-12.

178. Wilson, P.N. (1995) Getting the R&D act together. Landowning in Scotland, Autumn 1995. pp. 12-13.

179. Wilson, P.N. (1996) Land: Air: Sea – Sustainability through one department. J.Foundation Sci & Tech, Feb. 1996. 10-13.

180. Anon. (1996) 1996 Research Assessment Exercise – The Outcome. Higher Education Funding Council Circular ref RAE96 1/96. Bristol. 180 pp.

181. Webster, C.C. & Wilson, P.N. (1998) Agriculture in the Tropics. 3rd edn. Blackwell Science.
182. Wilson, P.N. Bovine Somatatrophin. In Engineering Genesis (Eds.) D. & A. Bruce. Earthscan Publications, London. pp. 55-59.
183. Wilson, P.N. (2000) A Tale of Two Trusts. The Memoir Club, Durham. 234 pp.

Index